BONZO'S WAR

CLARE CAMPBELL

Constable & Robinson Ltd
55–56 Russell Square
London WC1B 4HP
www.constablerobinson.com

First published in the UK by Corsair,
an imprint of Constable & Robinson, 2013

Written with Christy Campbell

A copy of the British Library Cataloguing in Publication data is available from the British Library

ISBN 978-1-47210-679-7 (hardback)
ISBN 978-1-47210-687-2 (ebook)

1 3 5 7 9 10 8 6 4 2

Printed and bound in the UK

BONZO'S WAR

Also by Clare Campbell

Out of It: How Cocaine Killed My Brother

Tokyo Hostess: Inside the Shocking World of Tokyo Nightclub Hostessing

To Fergus and Luis, and all those that went before

Contents

Contents

PART THREE: PETS SEE IT THROUGH

PART FOUR: PETS TRIUMPHANT

Author's Note

When I was eight years old I was horrified to discover that my uncle 'had killed a dog'.

Eavesdropping on my parents' conversation in our south London home one evening, I heard my mother tell my father how distressed her twin sister, Lena, had been at the very start of the Second World War. This was not as a result of fears for herself or her young son at the thought of the approaching conflict, but rather because her husband, Ernest, had decided to have their beloved pet dog 'Paddy' destroyed.

From what I could gather, Paddy had been a very nice Wire Haired Fox Terrier. I had never known him; it was all several decades earlier. But I do remember a fading snap in our family photo album of an agreeable, eager-looking dog with magnificently tightly curled fur bouncing along by my aunt's side. He had reminded me of the Tri-ang toy push-along dog that I myself had loved so much as a child.

Now he was long gone. But I pricked up my ears and wanted to know more. With some reluctance, my mother told me.

The way she told it, each evening my Aunt Lena would walk Paddy across the common to the suburban station to meet Ernest off his train, the dog jumping up to greet him with joy. It was summer 1939; war was coming. Everybody sensed it, even if the pets of the southwest London suburbs were as yet blissfully unaware of Herr Hitler's intentions.

Then, when the Invasion of Poland was at last announced on the BBC News, my mother told me that Uncle Ernest had suddenly decreed that Paddy must go.

Always a very 'rational' man and totally lacking in sentiment, he took the dog from my aunt's arms the following day (2 September 1939) and went out the front door, I assume, with Paddy happily beside him on a lead. The next day, a Sunday, the sirens sounded and Britain really was at war.

My aunt never saw Paddy again. I was horrified. I thought of my uncle, otherwise a twinkly-eyed, kindly man of whom I was very fond, and decided then and there that he must be a monster. 'Well, it was the war, darling,' my mother explained unconvincingly. 'Food was going to be rationed, everybody thought so, and your Uncle Ernest decided Paddy was one more mouth to feed. He wasn't the only one who thought like that at the time.'

For every person who thought like my uncle, I am sure that there were dozens for whom pets were like members of the family, and only slightly less dear to them than their own children. What on earth should they do if, as everyone expected, the war began with a cataclysm from the air?

As I would discover, it was a scene repeated in thousands of loving homes – weeping children, sobbing mothers, stern fathers saying it was the only thing to do. That it was the *kindest* thing to do.

It was all based on a false assumption: that mass bombing of cities with gas and high explosive would very soon follow the outbreak of war. A general panic stalked pet lovers. Would their animals suffer terribly? Would they become hysterical and run wild at the sound of sirens and explosions, their bodies burned or contaminated with mustard gas or whatever other horrors were coming? If they survived the opening onslaught, could they even be fed?

It had been in the papers, broadcast on the BBC – 'It really is kindest to have them destroyed,' said the man from the Ministry of Home Security, spoken in a soothing announcement from deepest officialdom made in the treacliest tones of Mr Cholmondley-Warner (I can only imagine that's how it was).

The result was a catastrophe. Actually it was all an accident, wasn't it?

And thus it happened that in those first days of war, many, many thousands of cats and dogs were shooed out into the street, dumped in the woods or taken to the vet or animal welfare clinic for some sort of kindlier end to be inflicted. It was a national tragedy, a heart-breaking pet holocaust that still haunts animal lovers like me. And it had touched my family directly.

'There has always been a rumour that something like that happened,' the archivist of the Royal Society for the Prevention of Cruelty to Animals (RSPCA) told me while I was on the long and varied research trail for this book. Well, it did happen – and it's all there in the Society's own archives – and in those of other animal welfare charities – and it's recorded in detail in the official files of HM Government. I know who I blame …

Introduction

So, what was it like to be a pet during the Second World War? What happened to domestic dogs and cats when bombs began to fall and food went on the ration? And what happened to the animals at the zoo?

Living in a house full of cats and history books I began to wonder. Unlike medal-winning 'war animals' with memorials and biographies galore, at first I could find nothing on the ordinary pets of ordinary people, and the humble working animals: the milkman's horse rather than the war horse. Then, little by little, the story opened up. What follows concerns the civilian animal experience of war and especially the pets of embattled Britain.

Since 2004, those animals who *officially* served Great Britain have had their own monument in Hyde Park, London: the 'Animals in War Memorial',[i] dedicated, as it

i Britain's two-and-a-half million (my estimate) companion-animal war dead have no physical memorial. Hyde Park is for service animals only, as is the more modest memorial unveiled in 2011 at Morley, Leeds, dedicated 'To all animals who have died serving their country alongside British troops.' The Civil Defence memorial at the National Arboretum, Staffordshire, honours the '1939–45 animal friends who served with such loyalty and bravery.' The Purple Poppy campaign launched by the pressure group Animal Aid in 2006 commemorates animals expended by armed forces in war and used in weapon experiments.

proclaims, to 'all the animals that served and died alongside British and Allied forces in wars and campaigns throughout time'. It bears a second inscription – 'they had no choice'.

Well, neither did the war pets: the companion animals that often had more to fear from their owners than anyone else, it would turn out. Cats and dogs had no choice when the bombs began to fall (other perhaps than to run), although plenty of choices would be made for them.

Pets, on the whole, do not write their memoirs. To find out what happened I had to look for accounts of those who entwined their lives with their pets because they loved them especially or because they sought to advance the wellbeing of animals generally.

I found them in abundance. And what extraordinary stories they were – of animals under fire in the Blitz, of evacuated, blacked-out, foodless and homeless pets, of brave cats and intrepid dogs who did not go barking mad at the first wail of sirens as everyone expected them to, but rather gave aid and comfort to humans.

And it is about how many humans, otherwise engaged in waging total war, did their utmost to comfort and save their animals. At least some of them did.

I looked at my own cats. Could I queue all day for a cod's head or boil up sheep's windpipe for hours on end? That took a deep devotion. Would I have smuggled them into an air raid shelter to be barked at by a horrid warden – or risked being taken to court for giving them a saucer of milk?

The PDSA stated in its post-war history: 'There is a field looking north from the Ilford Sanatorium which remains the officially recognised cemetery of some three quarters of a million cats and dogs.' After seventy years however, the site is unmarked and unacknowledged, a few cryptic mounds in scrubby wasteland on the banks of the River Roding.

Then there were the accidental animal combatants who became entangled by the fortunes of war, those who were left behind when everybody else had fled – like the Dunkirk dogs – and the masses of camp-following pets who often had to dodge officialdom or switch sides to stay alive. 'No mascot is as popular as one captured from the enemy,' it was said in the midst of war. And I imagine whoever wrote that knew what they meant.

Indeed many so-called 'regimental pets' had to change sides to survive at all. But who could condemn these furry collaborators with no understanding of the struggle in which they had been swept up?

And, as I discovered, Britain set out in 1942 to recruit an army of 6,000 dogs. They were pets loaned by their families for the duration. Many did not make it home. At the same time officials were secretly plotting the mass destruction of dogs and launched a hate campaign against cats. The wartime Archbishop of Canterbury would not allow the saying of prayers for animals because it was 'theologically inaccurate'. It's amazing that pets made it at all.

The war was won. And so the victor nation's animals had seen it through. Rationing would ease though not quite yet. In the glow of victory, the pets had done their bit by boosting morale – while in the background, Government officials had considered getting rid of them all in the cause of national survival. A bit like my Uncle Ernest, that part was not to be spoken of.

Knowing what I know now, I forgive him.

A Note on the Animal Welfare Charities

Britain historically has had far more animal welfare charities than political parties. And they fought like cats. As an official who dealt with them wrote, 'Anyone who knows anything about charities knows that, in their relations with each other, charitableness is their least conspicuous virtue'.

The Nazis simply abolished the lot when they seized power in 1933 and imposed a single state institution – the 'Reichstierschutzbund'. A British attempt on the eve of war to create something similar, the 'National Air Raid Precautions Animals Committee' – a kind of Dad's (and Mum's) Army for pets – had grand ideas but would struggle to make an impact. To understand how the charities worked with pets and how they recorded the experience, it is important to understand their contradictions and rivalries. A lot of small-animal welfare historically consisted of catching as many street animals as possible (strays, female kittens, mongrels) and killing them with chloroform in a 'lethal chamber'. Methods of destruction became more scientific but the conflict between the sentimentalists and the realists was always intense.

Nobody should have been under any illusions. As a pre-war Our Dumb Friends' League (ODFL) report stated bluntly: '[Our cat] shelters should not be confused with "homes" for cats. Their special object is to rid the streets of these unfortunate animals, not to keep them for a lengthy period without prospect of future homes.'

Much of the story of pets at war is of animals dodging bombs and falling masonry only to be 'mercifully put to sleep' by their rescuers. Certain charities gloss over this aspect of their wartime record in their published histories.[ii]

From their beginnings, the charities were prone to internal rebellions and provincial breakaways. They competed fiercely for donations and legacies, launching stunts especially in wartime, to loosen the patriotic animal lover's purse strings. They all scrabbled for aristocratic and royal patronage, which was inconvenient when it came to internal

ii In 2013 the RPSCA said of its wartime record, 'of the animals rescued from bombed sites during this year [1940], 10,100 pets sadly had to be put to sleep because of the extent of their injuries. But 5,940 animals survived and were successfully rehomed.'

Its archived end-1940 reports say in contrast, 'In one month alone last year [September] 10,100 household pets were humanely destroyed, 5,490 were rescued from bombed premises, fed for a time, boarded, or [432] provided with new homes.' The 1941 report records its inspectors dealing with 42,095 animal 'victims of the war' of which '50 per cent [were given] a painless end'.

The PDSA states in 2013 with justifiable pride that its brave 'Animal Rescue squads helped to save and treat over a quarter of a million pets buried and injured by debris during the Blitz.' Its 1947 history meanwhile accounts for the 'treatment' of almost 3.68 million animals in the six war years ('including lethalling'), while this number excludes the mass panic-killing of 1939–40. Not all, of course, were due to enemy action or wartime circumstances and not all were destroyed.

Battersea records '145,000 dogs [passing] through during the course of the war', while the Metropolitan Police reports the destruction of 77,217 of them by its south London contractors and a further 9,236 by the ODFL North London Dogs Home in the period 1939–44. 'There is no statutory requirement on the police as regards cats,' it was noted, 'no figures are therefore available.'

ideological disputes over fox hunting. A duchess or two was essential at fund-raising functions, joined increasingly by film and radio stars. Derided by cynical journalists as an activity for social climbing, middle-class ladies, the animal welfare champions, were much more complicated than that. Some were intensely political (on the left and right). Others just loved their pets, and many worked alone. Wherever bombs and rockets fell, there would soon be women with baskets doing what they thought was right.

Mr Keith Robinson, the secretary of Our Dumb Friends' League, admitted in a 1938 interview that 'anyone who devotes his life to fighting cruelty in animals is bound to be a bit of a crank'. Robinson revealed that most of his funding came from the donations of spinsters and was no more than a few shillings at a time. He himself lived with seven cats.

A wartime account described: 'One East End woman, who gave her bedroom to *the cause*, would herself search miles in a night with a box on wheels like the slum child's cart looking for injured animals.' But what could she really do to help them?

As I read the stories of selfless volunteers, of midnight feeders of ferals and those who listened for cries beneath the rubble, I realized they drew their fortitude from an earlier tradition of concern for animals, one with largely a female face.

The idea of 'pets' had risen with the late-nineteenth-century urban middle class (with Britain, the US and Germany leading the way) and now, in later life, it was their children who championed pets in their time of trial. Without mad cat ladies (and some were really mad), there would have been a much-diminished 'animal welfare' culture to meet this new emergency.

Even as late as 1939, the 'animal welfare movement' remained imbued by the spirit of its founders – many of them the impassioned women who, four decades before, had found another channel for their suffragette political energies in animals: *The Cause*. But under the renewed shadow of war, they were seen as eccentric Edwardian relics.

Women like Nina, Duchess of Hamilton (derided as a 'crank' by MI5), the Swedish aristocrat Louise ('Lizzy') Lind-af-Hageby and Alsatian breeder Mrs Margaret Griffin, who took her amazing rescue dogs into action during the V2 rocket campaign of 1944–45.

There were some remarkable chaps as well, like Major James Baldwin, the Alsatian wolf-dog champion, Mr Albert Steward, who lived in a house full of cats in Slough and campaigned without cease for wartime felines, Captain T. C. Colthurst, 'Animal Guard Number One', Mr Bernard Woolley, who turned a Lancashire cinema into a dog rescue centre, and Edward Bridges Webb, the inspired pet-populist of the People's Dispensary for Sick Animals of the Poor (PDSA), who invented the 'Mascot Club'.

The veterinary profession meanwhile remained overwhelmingly male, as were the executive officers of the charities, with lots of peppery former 1914–18 officers keeping everyone on their toes.

By the coming of war, routine cruelty to animals in plain sight had become rare. Working horses in cities had been largely mechanized away but there were new concerns. Pit ponies and performing animals generated strong feelings, as did Government poison gas and weapon experiments. Vivisection, vaccination and zoos always excited the ultras. The charities each had their own priorities. They further had deep experience, both political and practical, from the First World War and from the social hardships of the thirties.

The RSPCA (founded in 1824) was the oldest and richest charity of all. In 1932, led by its chairman, the Conservative

MP Sir Robert Gower, its Council had crushed an anti-hunting, anti-circus rebellion. The Society was interventionist and in many poorer people's eyes, its inspectors seemed to possess civil police powers and they were therefore suspicious of it. It had experience of mass, urban small-animal destruction in pre-war slum clearance programmes, killing an estimated 50,000 animals a year. In Birmingham, its staff had experimented using carbon monoxide from a car exhaust (and they themselves had been prosecuted for cruelty).

Through the war, the Society's annual reports and monthly magazine, *The Animal World*, were published unbroken. The Society had an ambitious foreign policy promoting animal welfare schemes in Poland, Finland (which were short-lived) and the Soviet Union.

Our Dumb Friends' League (founded in 1897) was not nearly so grand, and was less interested in prosecuting the cruel than the active promotion of kindness. The League had launched the 'Blue Cross Fund' to assist military horses during the Balkan Wars of 1912–13 and found support for its continuation in the Great War. It faced down a pre-war mutiny led by a countess over the alleged 'disgraceful condition' of the North London Dogs Home.

The plight of 'old war horses' sold on post-1918 to beastly locals in France, Belgium and Egypt proved a big fund-raiser right up to the eve of a new war (an equivalent German campaign in 1938 to give old horses 'war comrade medals' and 'free oats for poor farmers' found nine British Army horses that had fallen into German hands, still alive).

Money flowed in. Two years into the war, a former dustman willed a huge amount to the League to look after his cat, but it had inconveniently disappeared. However, the League promised to track it down and make it comfortable till the end of its days.

The League had drifted uncertainly to the radical wing of animal rights – with a 'political committee' which campaigned among other things for a rise in the legal status of cats ('currently the same as a weasel') and for the National Trust to ban fox hunting on its land. One activist watched by Special Branch was a member of the British Union of Fascists.

There was a big row in 1938 with a mass defection of aristocratic patrons when the secretary, a Mr E. Keith Robinson, said: 'We feel that people would get just as much fun from a drag hunt as they would from chasing a wretched little fox across the country.' The League published annual reports throughout the war.

The People's Dispensary for Sick Animals of the Poor, founded by Mrs Maria Dickin in the East End of London in 1917, was a relative newcomer – offering free medical care for lowlier pets and working animals. Its relationship with the veterinary profession was turbulent. However, its dispensaries, mobile 'caravans' and volunteers would be at the forefront of animal rescue when British cities were bombed.

In a mid-war publicity masterstroke, the PDSA founded both the Services Mascot Club and endowed the 'Dickin Medal' for gallantry, named for its founder. But it was for 'service' animals only. In fact most of the recipients were carrier pigeons of the National Pigeon Service. Winners of the Our Dumb Friends' League's 'Blue Cross' and the RSPCA's 'For Valour' medallion for animal bravery are less well known.

At the war's end Mrs Dickin wrote a prospectus for an 'Allied Forces Animals' War Memorial Fund' to remember those 'animals and birds who have suffered and died on active service in our time of terrible need'. She wanted a 'practical memorial in the shape of ten mobile dispensaries'.

But the charity's founder furthermore included what she called 'those civilian animals who shared with us the horrors of the raids'. As she said:

> The P.D.S.A. Rescue Squads were eye-witnesses of their misery, and know how they suffered. Like their comrades on active service, these animals frequently ignored their own danger to stand by and help their owners; often they struggled to find and help them in blazing buildings, when they themselves could so easily have escaped to safety alone. Truly they were faithful unto death.

It would be 'a tribute to the Unknown Animals who gave their lives in service for us, or were innocent victims in our war, not theirs'. She even made an analogy with the tomb of the Unknown Warrior in Westminster Abbey. It was not to be.

A year later, pets had vanished from the appeal. The memorial was to be for 'the thousands of animals and birds [which] have helped our victory'. The Unknown Animals would remain just that. Only those who died on 'active service' would be remembered – by plaques on the side of the shiny new caravans.

It was the same over fifty years on. In 1998 the PDSA's Trustees rejected an appeal by the then just-starting Animals in War Memorial Fund (its primary charitable aim, to 'promote the military efficiency of the armed forces of Her Majesty's Government') as being 'too far removed from our objects'.

But when approached again, a new Director General replied: 'If you could offer a guarantee that the sculptor could include an animal wearing a Dickin medal, I am sure it would influence the decision of the Council of

Management.' It was duly so. £10,000 was given to the service-animals-only memorial.

The organization published the *PDSA News* throughout.

The National Canine Defence League (NCDL) was founded in 1891 during the first Crufts Dog Show 'to protect dogs from ill-usage of every kind'. Throughout the war it published the lively *The Dogs Bulletin*, full of gossip and wheezes to boost the cause of dogs. It concerned itself greatly with refugee dogs, operated clinics and was very active in promoting Air Raid Precautions (ARP) for dogs, building dedicated canine air raid shelters in Kensington Gardens and Sutton Coldfield. Since the start, the League had been regarded as 'the Opposition' by the Dogs' Home, Battersea. On the eve of war, it was said that their secretaries 'could never work together'.

The Cats Protection League (CPL) was founded at a meeting held at Caxton Hall in London in 1927, under the chairmanship of Miss Jessy Wade. The first secretary was Mr Albert A. Steward and the headquarters were established at Prestbury Lodge, a sizeable house in Slough, gifted (as so many were) in a legacy, where he lived on the first floor surrounded by cats. Early achievements included the introduction of an elasticated collar for cats and 'the development of a simple cat door'. What a boon – and still flapping down the ages!

Its magazine, *The Cat*, is the indispensable source for the British feline view of the Second World War and I am grateful to the current editor, Francesca Watson, for permitting me access to wartime copies kept at the National Cat Centre in Ashdown Forest. Long may its work continue.

A Note on the Sources

Pets, on the whole, do not leave diaries, memoirs or letters. Fortunately for the author of this book, the British obsession with the domestic pet meant that there was an outpouring of words written by humans on the subject throughout the years of the Second World War.

The principle animal welfare charities of the period mentioned above have largely survived and I am grateful to their archivists. They published their own magazines (*The Cat*, *PDSA News*, *The Animal World*, *The Dogs Bulletin*, *The Animals' Defender* etc.) and annual reports, while the turbulent affairs of the semi-official umbrella organization, the National Air Raid Precaution Animals Committee (NARPAC), are amply recorded at the UK National Archives.

In spite of paper shortages, bombing, and evacuation, those splendid enthusiast publications such as *Our Dogs*, *Cat World*, *The Dog World*, *Fur and Feather*, *Bee Craft*, *Cage Birds*, *The Goat*, *Kennel Gazette* etc. kept going throughout the conflict with their own insightful reflections on total war. The *Tail-Wagger Magazine*, fabulously, featured articles contributed by pets.

Horse & Hound, *The Field*, *Farmers Weekly*, *Eggs*, *The Smallholder*, *The Veterinary Record* etc. take the story of wartime animals into a pastoral context. And wartime newspapers, local and national, had a passion for hero animal stories that continues undiminished. The archives of the Zoological Society of London, the Imperial War Museum Department of Documents and Mass-Observation are all pet-friendly.

The files of the Ministry of Agriculture and the Ministry of Food tell the alarming story of the dwindling food bowl – the issue, apart from the actions of their owners, which truly determined the fate of wartime animals. War Office and Air Ministry files contain the official story of Britain's war dogs, while Home Office files recount the amazing tale of the 1944–45 London rescue dogs.

A Note on the Naming of Pets

All pet names, – 'Dusty', 'Blackie', 'Little One', 'Teeny Weenie', 'Hitler' etc. – are as originally reported.

Bonzo was found by Mass-Observation to be one of the most popular dog names of 1941. It was still all the rage after the cartoon dog first drawn by Englishman George Studdy in 1922, which inspired a worldwide craze and the naming of a vast number of real-life Bonzos (including two pre-war winners of the *Daily Mirror* Brave Dog award). From 1929, there was a feline equivalent, 'Ooloo'.

Oo-Oo (*sic*) was a Maida Vale cat who came into Our Dumb Friends' League hands in unusual circumstances on the eve of war. I would like to have met Oo-Oo – in fact I am sure I have done. These two will be our guide to certain wartime events.

Abbreviations

ADL Animal Defence League
ARP Air Raid Precautions
ATS Auxiliary Territorial Service
AWDTS Army War Dog Training School
BEF British Expeditionary Force

BUF British Union of Fascists
CD Civil Defence
CMP Corps of Military Police
CPL Cats Protection League
LAPAVS London and Provincial Anti-Vivisection Society
MAFF Ministry of Agriculture and Fisheries
MAP Ministry of Aircraft Production
MFH Master of Fox Hounds
MFHA Master of Fox Hounds Association
M-O Mass-Observation
NARPAC National ARP Animals Committee
NCDL National Canine Defence League
NVMA National Veterinary Medical Association
ODFL Our Dumb Friends' League
PDSA People's Dispensary for Sick Animals
RAVC Royal Army Veterinary Corps
RCVS Royal College of Veterinary Science
RE Royal Engineers
RSPCA Royal Society for the Prevention of Cruelty to Animals
USAAF United States Army Air Force
VP Vulnerable Point
ZSL Zoological Society of London

Part One

PAWS IN OUR TIME

In Memoriam

Happy memories of Iola TW. 695778. Sweet, faithful friend, given sleep September 4th, 1939, to be saved suffering during the war. A short but happy life – 2 years 12 weeks. 'Forgive us, little Pal, you were too nervous to be sent away. Au revoir' Terribly missed by all at 6, St. Ives Road, Birkenhead. S. 66

Tail-Wagger Magazine, October 1939

Chapter 1
In Case of Emergency

Nobody could say they had not been warned. When news came on 1 September 1939 that Poland had been invaded only the most obtuse British pet owner could believe that, this time, the long-heralded war with Germany could be averted.

Look at the fuss there had been twelve months before. In the space of two frantic weeks, Mr Neville Chamberlain, the British Prime Minister, had flown to Germany three times to meet Herr Hitler and discuss the limits or otherwise of the Führer's territorial demands in Europe. His last destination had been Munich.

An animal-loving politician (and there were some such) wrote not long after the events: 'On 28 September 1938, men and women awoke with the idea that this would be the last day of peace and within, maybe, 24 hours London would be drenched with poison gas or reduced to ruins with high explosive or both.'

That was the day that Hitler's 'give-me-your-frontiers' ultimatum to poor little Czechoslovakia was due to expire. It really looked like this was it.

The night before, Chamberlain had made his 'far away country of which we know nothing speech' on the wireless, after which 150,000 better-off Londoners departed the bomb-menaced capital sharpish for Wales and the West Country in what was described as a 'continuous rush of cars'. In the great exodus some took their pets with them to find refuge in animal-friendly boarding houses. Some heartlessly left them behind, while others took more extreme measures.

The Our Dumb Friends' League's kennels at Shooters Hill, southeast London, reported plenty of panicky calls demanding a speedy solution to their pet problem. 'We urged all these enquirers not to be stampeded into premature action, but often without success,' stated the League's end-of-1938 report.

'Many were hastening into the comparative safety of the country in cars piled high with their belongings and only stopped long enough to hand over their dogs or cats to be kept until either the situation improved or the worst happened, in which case they were to be destroyed forthwith,' it continued.

There was a less drastic but expensive alternative. Some better-off, stick-it-out city dwellers responded to such press announcements as this, made on the 28th by the Bellmead Kennels at Haslemere in Surrey: 'Ensure the safety of your pets. They will be safe and happy in the quiet depths of the country – away from the horror of noise.'

How reassuring. As war clouds gathered over the Home Counties, there were kindly folk who were there to help. Mrs Barnes of Suntop Kennels, Tonbridge in Kent announced: 'Send your dogs to happiness and safety. Reduced charges during emergency. Train met.'

I would have gone to stay there myself.

And on the following day, when Hitler invited Mr Chamberlain for a final showdown at Munich, the patrician Mayfair Dogs Ltd of Curzon Street, London W1, wished 'to inform owners who would prefer to have their dogs and cats out of the way during the present emergency that they can accommodate them in suitable kennels in the country'.

The drama was not restricted to domestic pets. Other animals were in the firing line. The London Zoological Society added to the gloom by announcing on the 28th: 'In the event of war all poisonous snakes and spiders will be immediately killed. Should any large animals escape as a result of damage to their cages they will be shot. Men have been detailed for this eventuality.' The aquarium director, Mr E. G. Boulenger, would confide later that as the crisis deepened, he turned off the heat in the reptile house to induce death.

Crowds watching the comings and goings in Downing Street cheered every time a mysterious black cat strode across from the Old Treasury Building (*see* p.34) and sat on the doorstep. Was it good luck? A German tourist told a reporter that in her country the appearance of a black cat signified very bad tidings indeed.

When on the 30th Mr Chamberlain flew back to London with his piece of paper, there was general rejoicing. The ministerial feline was sent a parcel addressed to the 'Black Cat No. 10 Downing Street'. It contained two Dover soles sent by an anonymous well-wisher. The heat in the Regent's Park reptile house was turned on again, just in time. For the nation's pets it was 'Paws in Our Time'.

What happened during 'Munich' is important towards understanding the enormity of events a year later. That modern war meant the targeting of civilians had long been clear. The bomber aeroplane could apparently bring carnage to anyone's front door, as had recently been made evident in China and Spain.

For months animal welfare groups had already been planning for such a catastrophe descending on Britain.

Sir Robert Gower, Tory MP and chairman of the RSPCA, told a newspaper in May 1938: 'The experience of Spain [the attack on the Basque city of Guernica] sadly shows the lethality of air attack and many animals would have to be put out of their suffering. Cats, as in many other matters [what did he mean?], are a particularly difficult problem.'

A 'special, humane cat catcher is being devised,' he announced.

The resulting 'Cat Grasper' (price 4*s.* 6*d.* including postage) was featured in its advice pamphlet *Animals in Air Raids* published that summer, a horrible contraption like a fishing rod with a running noose, along with an illustration of how to use it – 'to secure and shoot a cat that has been contaminated with mustard gas'. The use of clumsy gloves with a Webley & Scott captive-bolt pistol and a struggling animal was impractical – far better to use the grasper, it said. Readers were assured that the 'gas-contaminated' cat in the accompanying grisly photograph was already dead. It is hard to imagine that anyone could seriously regard such a medieval instrument of torture as in any way humane.

'Air Raid Precautions', as it came to be soothingly called, was all anyone would talk about that summer. And the prospect of mass evacuation from endangered cities. What will you do? Where will you go? What about the children? Do you have pets? Better think ahead.

*

In July 1938, the civil-servant-turned-MP Sir John Anderson, a 'po-faced rightwing bureaucrat', as he would later be described, had delivered a comprehensive report on what could or could not be done. The pre-fab, family-size air raid shelter, devised in 1938 to be sunk in the garden, was named after him. Officials at the Home Office set out to try and turn the recommendations into practical plans – but what about pets? A tiny 'departmental committee' began to study the issue.

Then in September had come 'The Crisis'. Munich caught animal welfare charities as and where they were in the general panic. Their immediate concerns were, it might seem, for their own prestige.

The RSPCA, for example, reported in October that in the crisis week it had been 'inundated' with 'what to do?' pleas from pet owners. 'When the good news came through, one could not help reflecting with pardonable pride that it was to the RSPCA that the masses had turned for practical advice,' said its journal, *The Animal World*.

'During the 48 hours in September when war was in the balance, the society's headquarters received 3,000 telephone calls, some 700 personal inquiries regarding the welfare of animals if hostilities came, and over 10,000 copies were distributed of the pamphlet which the Society alone had had the foresight to prepare in regard to the care of animals in war,' chairman Sir Robert Gower MP would later boast on a fund-raising platform.

The Society, very practically, had had discreet pre-war discussions with German counterparts, where an animals-in-air-raids-protection organization, the Luftschutz-Veterinärdienst, had existed since 1934. Its primary manual, *Luftschutz der Tiere*, drafted by Generaloberveterinär Professor Dr Claus Eduard Richters, concerned farm animals more than pets but expressed views with which British pet lovers would sympathize. 'In the animal we see not only something useful and valued,' wrote the military vet, 'but also a creature with feelings of its own, a helpful, loyal comrade, for whose welfare we feel just as responsible as for that of our fellow human beings.'

Non-Germans and their animals meanwhile were being evicted from their Czech homelands under the terms agreed at Munich, the Society could report. But the state of chaos in the capital, Prague, meant they had received no reply from their Czech counterpart on the fate of the Sudetenland pets. 'Appawsment' had abandoned them to their fate (it did not say that).

The National Canine Defence League reported the British dog lovers' view:

> The last few days of September, when it seemed that nothing but a miracle could save our country from being plunged into the most devastating conflict of all time, will long remain in our memories. Here at the League's headquarters, when the war clouds were at their blackest, one bright ray illumined the gloom, and that was the realisation that British dog-lovers were more concerned for their canine friends' safety than their own.

How true. The veteran anti-vivisectionist Louise Lind-af-Hageby, stalwart of the Animal Defence Society, universal 'peace' and animal activist since the 1900s, recorded: 'During the war-crisis of September, 1938, the Council of the Society had considered ways and means of helping in the event of cities being bombed. The possibility of large kennels in country districts and of utilising the Society's two motor caravans had been discussed. The Council feared that large numbers of dogs and cats would be destroyed if evacuation became necessary.'

Her co-activist and neighbour, Nina, Duchess of Hamilton, who sheltered two urban foxes, 'Vicky' and 'Patrick', in her St John's Wood mansion, magnanimously declared the splendid Hamilton country estate at Ferne in Wiltshire to be a safe haven. As she wrote later:

> During the September crisis Mrs Freeman sent six cats. When the crisis was over they were returned to her from Ferne Sanctuary and she wrote as follows: 'my six pets arrived here on Friday evening from Ferne, all of them in very beautiful condition and, incidentally in the highest spirits. They have benefitted in a remarkable way from the care and surroundings they have been privileged to enjoy, I should like to express my deepest gratitude'.

So the crisis turned out alright for some posh cats. In fact Ferne would be opening its doors to plenty more animals before too long, posh and not so posh. Mrs Freeman's cats would be back.

It was a general fear of poison gas that gave the Munich panic its particular edge. And there were real fears in Government that Germany might use animal diseases, especially anthrax, as a weapon.

As the National Canine Defence League found:

> All through the week of throbbing tension that preceded the Munich Conference, the League's telephones had scarcely an idle moment. Where can I get a gas mask for my dog? What do you advise me to do with my dog in the event of air raids? Where can I find accommodation for my dog in the country?

The RSPCA could report 'another enquirer caused something of a diversion by asking whether she could be supplied a gas mask for bees. The enquirer was animated by genuine concern.'

Cat World magazine reported on 1 October: 'We have had numerous letters and telephone calls from anxious readers asking for information regarding the safety of their cats should this country be subject to air raids and poison gas attacks. The RSPCA suggests sending your pets away to friends, finding refuge in a gas-proof room, but it is not possible to obtain gas masks for cats.'

The Cats Protection League was no less anxious. As its journal, *The Cat*, had reported earlier in the summer: 'One of our members asks what are the air raid precautions we suggest for cats and adds: "I should hate to have them killed unnecessarily".'

Just so. The League had asked the Whitehall authorities what plans were being made for felines and evidently did not get much of a reply. *The Cat*'s editor reported: 'Until we know for certain what would be the special arrangements we must NOT depend on being able to take our cats to Public Air Raid Shelters. One of the cold facts is that human beings will be considered first.'

And the humane and kindly columnist for *The Cat*, Mr Albert Steward, was right, of course. He could also report meanwhile that there was no practical way to protect cats from gas attack. 'Those of us who have seen the sufferings of homeless cats on the continent would not hesitate to make the decision that would make their cats safe from the horrors of war,' he wrote. 'Kill them' was what he meant.

What on earth to do? Many pet lovers decided to say goodbye. As the National Canine Defence League reported at the end of that dismal year: 'One feature of the crisis which should be placed on record is that many dogs were brought to the NCDL clinics to be put to sleep. Our clinic superintendents declined to destroy these healthy animals, advising the owners to wait and see. Dozens of dogs were thus reprieved, and are still enjoying life to-day.'

Six weeks after Munich came a brutal new twist. The 'Kristallnacht' pogrom in German and Austrian cities set a new flood of Jewish refugees to flight. Some of them had pets. Frau Weingarten of Vienna had her cats' quarantine fees paid by the Cats Protection League after a national newspaper appeal. Our Dumb Friends' League also knew what to do. 'The League felt that this country, which prides itself on its love for animals, could not lower its prestige in the eyes of so many foreigners by deliberately killing their pets,' it reported. 'Not only was it a national, but an international duty to save them.' And so, with much anti-foreigner grumbling from the sidelines, their quarantine kennels were opened to asylum-seeking dogs.

The Dogs Bulletin reported: 'As one penniless refugee from Vienna said to the NCDL, as her St Bernard dog was accommodated gratis, "I shall always be indebted to you".' But there were more warnings that refugee dogs, and those left behind by evacuees, might fall prey to vivisectors.

Pets faced danger everywhere.

Chapter 2
It Really is Kindest ...

The 'Peace in Our Time' jubilation after Munich did not last long for the last Republican strongholds in Spain, Barcelona, then Madrid, were on the brink of falling. The plight of refugees trudging with their animals across the Pyrenees from Catalonia into France in February 1939 excited the RSPCA enough for them to send the veteran colonial vet, Colonel Robert John Stordy, to Perpignan on the border to see what could be done. He found the French Army had already 'commandeered healthy horses and mules and arranged for the humane destruction' of the rest. There was little more to do but come home.

Colonel Stordy had already drafted a memorandum for fellow vets and the Scottish SPCA on 'animals in a national emergency'. He had suggested in January that the 'experience of Spain' and its bombed cities showed that high explosive and incendiary attack was more likely than gas – although owners of cats and dogs should practise wearing masks so 'their pets would become used to the ghoulish appearance of their owners' and recognize their muffled voices. 'Should hostilities eventuate, there is no doubt that the abandoned dog and cat will stand in the greatest need of concerted action to put an end to their suffering,' he wrote.

He even covered school pets, 'such as rabbits, guinea pigs, etc., housed on school premises for the primary training of children in the care of, and kindliness to animals, [which] will have to remain in the charge of the care-takers of their respective institutions'.

Crufts Dog Show for 1939 meanwhile, held at the Agricultural Hall, Islington, on two chilly February days, attracted 'thousands of dogs'. Among the many highlights were large contingents of Pekingese, Wire Haired Fox Terriers (the super-fashionable dog, like my aunt's poor doomed Paddy, of the time), Welsh Corgis and Chow Chows.

'Ivor of Dunkerque', the Wire Haired Dachshund exhibited by Air Vice-Marshal Sir Charles Lambe (famed as the 'first Englishman to own a German Wolf Hound') proved a champion. The Zoological Society of London exhibited a Husky. Tibetan Mastiffs and Basenjis from the Belgian Congo provided an exotic touch.

There was a whiff of decadence as war clouds gathered beyond the judging ring. The *Daily Mail*'s Patrick Murphy witnessed 'dogs dressed in Angora wool jumpers with Magyar sleeves' as well as 'quite sizeable dogs wearing trousers'.

Best in Show was Mr H. S. Lloyd's Cocker Spaniel 'Exquisite Model of Ware' for the second year running. More would be heard of Mr Lloyd when the nation's pets faced a sterner test than even the judges' bench at Crufts.

On the other side of town, the Dumb Friends' League's kennels at Shooters Hill, Blackheath, run by Colonel and Mrs W. M. Burden, were home to an increasing number of refugee dogs – such as 'Barbara' who featured in the 1939 report, the pet of two Jewish children, Elizabeth and George Mayer. They had written this heart-rending plea:

> We two children have got a permit to emigrate to England. But we can't come because we must leave our dog 'Barbara' alone. It was a present to us for our birthdays two years ago. At that time Barbara the dog was only two weeks old. We have brought her up and she has been accustomed to us in such a way that she would fret if we would leave her.
>
> But also we are not able to live without her. We know that Barbara must be quarantined for the first half of the year but we have not sufficient money to pay for her keep. We are very poor refugee children and we love Barbara, who would die if we abandoned her. Therefore we ask you if you would keep her for six months free of charge or if you have a good friend who would pay the expenses. We can show the papers that the ancestors of the dog were born in Scotland.

The League was proud to announce it was meeting the costs of boarding Barbara, as well as 'all these foreign dogs' in their statutory six-month quarantine.

The Canine Defence League could not have done more to extend the paw of friendship. It dramatically reported the success of its Quarantine Fund for Refugee Dogs appeal. 'As a result of the help thus given, the future of the pitiful canine outcasts from Central Europe, for whose care the League has so far assumed responsibility, is now assured,' it was reported in *The Dogs Bulletin*. It really was terribly moving:

> Many of these refugee dogs' masters and mistresses arrive in this country with little else but the clothes they wear and it speaks volumes for their affection for their canine friends that in some instances they have denied themselves necessaries in order to ensure that their dogs

> should breathe the free air of England. Seldom have we of the League derived more pleasure from aiding downcast dog-lovers than we have in extending a helping hand to these victims of racial and religious persecution.

There were also feline refugees from Nazism. *The Cat*, a little later in the drama, highlighted the case of 'Mischi' – 'a really lovely striped tabby with large intelligent eyes and long, graceful lines. She has a large vocabulary in two languages, one of them Czechoslovakian, although she is an "England" cat, and instantly replies to certain words of interest, such as "meat", etc.'

Mischi, now living in Slough, was pregnant. Her owner was looking for a good home for 'a son of this beautiful and intelligent cat'. Give him to me!

As the German Army marched into the capital of Mischi's homeland, Prague, on 16 March, the national mood darkened. Sir John Anderson, who had entered the Cabinet as Lord Privy Seal in October 1938, intensified defensive preparations. Maybe this time the nation's pets would get some proper consideration.

Animal lovers could reflect meanwhile that as Europe was rushing towards the precipice, at the nation's helm was someone who understood the natural world. The Prime Minister, it was noted, was a keen birdwatcher. 'His morning bird walk in St James's Park accompanied by his wife must be of considerable value in providing physical and mental refreshment before facing the day's duties with the Cabinet,' reported *Animals and Zoo Magazine*.

Under the attention-grabbing headline 'Neville Chamberlain Moth Hunter', the magazine recorded the PM finding a wood leopard moth in 'the venerable garden of Number 10'.

'It is certainly no small encouragement to natural history

in these difficult times that the Prime Minister has found time to collect moths as well as to watch birds and fish,' said its correspondent.

But after Prague, war really did seem inevitable. In early April, Whitehall officials met to discuss what to do with domestic pets in case of war – but not yet the larger animals. 'Cattle and sheep present little problem in the Metropolitan Police area,' it was noted at a meeting of the provisional committee set up to examine what was now being called an 'ARP for Animals Service'. The Commissioner of the Metropolitan Police, Sir Philip Game, got directly involved.

Just what it was supposed to do would depend on whether the Government ordered 'the compulsory slaughter of all animals not of economic value' according to an internal discussion paper (the proposal is marked 'suggest delete' in the margin). That meant the mass destruction of pets.

The British Bee-Keepers Association meanwhile complained that no one in authority was considering the position of bees, should war break out. The evacuation of hives from the suburbs of larger cities should be considered, 'because in an air raid, a badly disrupted apiary could increase panic and hinder rescue,' it was noted in the journal *Bee Craft*. That was not to be the last word on bees.

The RSPCA provided a census of the imperial capital's non-human population. 'Besides the 40,000 working horses in the metropolis there were also 18,000 pigs, 9,000 sheep, 6,000 head of cattle, 400,000 dogs, and approximately 1,500,000 cats,' announced Sir Robert Gower at a conference in May. 'The public looks to the Society to see that this vast army of animals has adequate protection.'

Many of those working horses were employed by railway companies, breweries, dairies and borough councils, but many more were the animals of the poor, living in tumbledown backyard stables in the sort of conditions that Maria Dickin had found so distressing, twenty years before. If cities were evacuated or bombed, how should they be looked after?

The Home Office discussions progressed through the spring. The National Veterinary Medical Association (NVMA) gave their views. The RSPCA was brought in, plus the chairman of the Dogs' Home, Battersea, Sir Charles Hardinge. Mr H. E. Bywater, chief veterinary officer of the County Borough of West Ham on the eastern edge of the metropolis, who was organizing an experimental local defence scheme, was co-opted.

It was agreed that some sort of official advice to the public should be drafted. The initial plan was for a regional network of vets to work with local authority ARP services, but when in March the 'various animal welfare societies' were at last consulted after months of being ignored, they fell over each other to get aboard.

The man doing the consulting was Mr Christopher Pulling, barrister, chronicler of the English music hall, connoisseur of detective fiction and career civil-servant (he would be Senior Assistant Secretary at Scotland Yard for thirty-five years). The intervention of this exotic figure on behalf of pets would have unlooked-for consequences.

'The RSPCA, People's Dispensary for Sick Animals, Our Dumb Friends' League, the Dogs' Home, Battersea and National Canine Defence League all have plentiful funds and are well situated with ample premises,' Mr Pulling noted in April. 'One is well aware of the complications of the animal defence societies' internecine politics, but I am assured by the secretaries of most that they are anxious to offer their full resources with no charge on public funds,' so he informed the Commissioner. He thought a strong, independent chairman would be necessary if they somehow agreed to co-operate.

Meanwhile he was developing a scheme of his own. To get away from danger, 'people with cars might be able to take their pets with them,' he said, 'but there are others who will not.' It was a statement of the obvious. Animal welfare societies should not only be responsible for the destruction of pets but also for the evacuation of animals in 'good condition' to suitable rural refuges, he proposed.

Nevertheless a large number of strays was inevitable. Starving or injured cats and dogs should be straightaway put down by police. Where London strays in 'good condition' came into police hands, they should be sent to the Dogs' Home, Battersea in the usual way (who would decide what was 'good' or not?) with the statutory retention period under the Animals Act, 1911, before destruction – in case anyone turned up to claim them. But the period of grace would be reduced if necessary by an Emergency Defence Regulation from seven days to three.

The Home Office animals-at-war advice pamphlet was progressing. It was a well-meaning mix of technical information about the effects of high explosive and various gases – 'that might be employed in a future war' – and the mechanics of despatching creatures who might be 'incurably injured'.

Its basis was Colonel Stordy's memorandum of January, with a dash of Professor Richter's *Luftschutz* manual, with amendments made by interested parties – including this inserted at a meeting of the 'drafting sub-committee' in April at the suggestion of Mr Pulling:[1]

1 Pulling told the Commissioner, 'I attended a meeting of the HO (ARP) Animals Committee on 5 April [1939] and was able to secure the inclusion of a paragraph advising owners to make up their minds in advance whether they wanted their pets evacuated, and if not, whether they wanted them destroyed.'

> Dogs and cats and other pets must be considered the personal responsibility of their owners. These animals will be prohibited from entering the shelters provided for public use. Owners should make up their minds whether they can take away their dog or cat themselves. If this is impossible, they should decide whether the animal is best destroyed or evacuated to the care of friends in the country.

The destruction imperative was repeated at the end – just before the appendix on how to use a captive-bolt pistol. It said:

> When an owner has been unable to send his dog or cat to a safe area, he should consider the advisability of having it painlessly destroyed. During an emergency there might be large numbers of animals wounded, gassed or driven frantic with fear, and destruction would then have to be *enforced* [author's italics] by the responsible authority for the protection of the public.

This was the primary fear that would now drive policy, the prospect of frenzied hordes of gas-contaminated cats and dogs swarming through burning cities. And what about those school pets? Instead of being cosily tended by caretakers such as Mr Stordy proposed, all those guinea-pigs, rabbits, etc. were not to be 'destroyed unless they can be evacuated in advance,' stated the manual published in early July as *ARP Handbook No. 12* – 'Air Raid Precautions for Animals', available from His Majesty's Stationery Office, price 3*d*.

Meanwhile, a grand conference of what were described as 'the first class animal societies' was held in London on 22 June (the exclusion of the 'lesser' ones caused outrage) under Sir

John Anderson's political patronage. Would 'an extension of their peace time activities' be enough to cope with pets in war? Could they 'pool their resources' under some new organization? A highly revealing discussion resulted, reported to the Commissioner by Christopher Pulling.

The West Ham vet, H. E. Bywater, suggested 'evacuation would only be the fringe of the problem'. In his borough alone there were '20,000 dogs and 60,000 cats'.

'The main problem would be the disposal of these pets,' he said, 'by industrial concerns [which] should make use of all by-products possible.'

Mr Edward Bridges Webb, secretary of the PDSA, agreed with this starkly utilitarian approach, suggesting that 'carcasses, once collected, be turned to profitable use', adding that his remarks referred to 'small animals only', as the minutes of the meeting recorded.

Meanwhile Mrs Beauchamp Tufnell of Our Dumb Friends' League 'thought the outbreak of war would be so sudden, evacuation would not be possible'.

There was general agreement that the outbreak of war would mean the mass killing of pets, even if not yet by Government compulsion. Were the means adequate? Both Our Dumb Friends' League and the Dogs' Home, Battersea were contractors to the Metropolitan Police in the matter of stray dogs, it was noted, with humane destruction as their main task. Indeed Battersea did so on an industrial scale.

The League used electricity, as did Battersea, which was capable of killing 100 dogs an hour with their very modern 'electrothanaters' on two sites (the second was in Bow, in the East End). Their vans could carry twenty dogs each. Each Canine Defence League clinic could deal with fifty dogs and thirty cats an hour by electrocution, chloroform or hydrocyanic acid injection. The RSPCA had fifty-two 'cat and dog lethalling centres' in London alone and more in provincial cities. The PDSA were proposing an evacuation and registration scheme with 'two kinds of [identity] disc, one to say the animal may be destroyed. They disposed of fifty 'Temple Cox Captive Bolt Humane Killers' with 150 staff trained to use them, the ministry was told.

The death chambers and captive-bolt pistols were ready. Mr Arthur Moss of the RPSCA gloomily pronounced that the 'primary task' for them all would be the destruction of animals and proposed that his inspectors be granted three months' exemption from call-up so that 'they could train persons in the use of the humane killer'. Lorries and lifting gear would be needed for the collection of carcasses and 'four-pronged forks for small animals'.

To dispose of the corpses, the firm of Harrison, Barber & Co., slaughterers and fat renderers of Sugar House Lane, Stratford E15, was ready to do its bit. Already there had been preliminary discussions with the company chairman, Captain E. Upton. A schedule was drafted of economically useful by- products – soap, fats, glues, fertilizer, fur. It was noted that the 'voluntary societies carried out [destruction] work without charge, the profit lying in the disposal of the carcass'.

It was looking grim for pets.

Thus it was by early August 1939 the Home Office contingency animals committee had become transmuted into part of the nation's defences. It would have a grand title, the 'National Air Raid Precautions Animals Committee' (NARPAC) and bring together, it was to be hoped, the veterinary profession, the animal welfare charities and the Government in working harmony.

Its chairman would be a retired Ministry of Agriculture undersecretary, Mr E. H. Dale. The existing committee was disbanded, its only transferee being H. E. Bywater, who would serve as honorary treasurer. The chief executive officer (appointed on 19 July) would be Colonel Robert Stordy himself.[2]

The National Veterinary Medical Association's liaison officer would be its own newly appointed president, the flamboyant, monocle-wearing (he had lost an eye as a young vet when a horse kicked him) Henry Steele-Bodger. Also on the committee was Keith Robinson of Our Dumb Friends' League and the thrusting Edward Bridges Webb of the PDSA, both of them enthusiastic populists, eager to reach out to the public through whatever means. Both agreed (the other charities did not) that they were already well-funded enough not to need a Government grant although 'an appeal to the public should be made'. After all, that is what they were good at.

But should this organization be swift to save or eager to destroy? Mr Robinson suggested in early August that voluntary 'Animal ARP Wardens', as he called them, should register animals in their area. And he told the Ministry this:

> There must be no suggestion of the immediate destruction of small animals. If the public has learned that the Wardens are there and should war break out, their animals will be immediately looked after, there will be no need for wholesale destruction at every crisis.

Mr Bridges Webb designed a striking badge to be used on armbands and posters – and immediately copyrighted it in his own name.

2 The first candidate approached, former assistant commissioner of the Metropolitan Police, Sir Trevor Bigham, ruled himself out as he already felt 'appalled' by the number of cats and dogs in the capital and was of the view, 'the least the pet-lover can do is put them to sleep before the trouble begins'.

Who was going to pay for all this? Vets were expected not to charge fees but what about their expenses? Who should pay the rent and for the establishment of an air raid shelter in the basement of Gordon Square? What about protective clothing, helmets, gas decontamination equipment, vehicles? Would local authorities provide them? Colonel Stordy's expenses were considered. It was agreed he could charge the services of his existing chauffeur, a certain Mr Badger, and be granted a petrol allowance of 4*d.* per mile. From what fund had yet to be agreed.

With the war crisis deepening, it seemed better just to plunge on with 'a short term scheme' and worry about the details later. On 14 August NARPAC's headquarters were established alongside the existing NVMA's headquarters at No. 36 Gordon Square in leafy Bloomsbury, where everything was beaming goodwill, for the first few days at least.

Meanwhile those Home County kennels were back, cheerfully offering 'safe accommodation for refugee dogs' – but hurry, only a few places left. A lady in Headington, Oxford, advertised: 'Gas proof kennels for dogs. Cats extremely happy in special cottage.'

I feel sure they were. A Worcestershire aviary offered sanctuary for parrots too.

Some enterprising pet owners decided to create their own domestic shelters. Gas was still the biggest fear for humans and animals alike but most of the advice to pet owners in the event of chemical attack seemed absurd. 'The contaminated hair on cats should be cut out and carefully destroyed,' recommended *ARP News*, while 'goldfish mildly off colour may revive after a raid of a pinch or two of Epsom salts dropped in new water. Liver is a good pick me up for dogs gone off their meals.'

The PDSA's advice for bird-keepers was simple enough: 'Hang caged birds off the floor in a gas-proof room below the level of any gas light-jets. If no gas-proof room is available, completely wrap the cage in a wet blanket. (This method should only be used as a last resource.)'

Gas masks for dogs were considered. *The Veterinary Record* pointed out the difficulties: 'In the first place, to get the animal to tolerate the mask. Secondly, there are the various shapes and sizes of dogs' heads which would necessitate a large number of different sized masks, e.g., Pekingese, Terrier, Chow and Mastiff.'

'It is obviously impossible to fit a gas mask to a cat or dog,' it was noted. 'For if a dog is fitted with a mask he loses his sense of smell and with it his sense of direction and as a dog perspires through the mouth he cannot perspire freely. Even if a satisfactory mask could be devised the animal's body and feet would still be vulnerable.'

But humans must carry gas masks. This too brought pet-based dilemmas. *The Veterinary Record* again:

> As a last general hint, owners should not only practise wearing their gas masks themselves for their own sakes, but by so doing they will accustom their pets to their changed appearance and muffled voice.
>
> A striking example of the necessity for this precaution is afforded by the unfortunate experience of a lady – the possessor of a terrier – who put on her mask and went down on all fours to play with her pet. The dog promptly bolted and has not been seen since.

To protect small animals from gas, it had to be some sort of container. Mr C. H. Gaunt of the PDSA devised a 'gas-proof safety kennel' with a hinged and sprung floor so that 'every movement of the animal works a bellows that passes filtered air'. It was totally impractical.

Mr Frank Heaton, a Midlands inventor, advertised his patent gas-proof small animal shelter, its air refreshed by a bicycle pump – twenty to thirty strokes per hour. If you died so too presumably would your pet. It appeared in *The Tatler* magazine and in July 1939 attracted this letter of appreciation:

> I enclose two 'snaps' of my Scotties in their gas kennels, which I've had very little trouble in getting them used to. One dog is particularly shy and nervous. I overcome this by always putting his favourite toy inside just out of his reach, so that he got quite used to going in and out. When I came to release him, he was quite happily curled up asleep.
>
> The other Scottie growls at him if he tries to get into his barrel! The younger dog likes to be inside his barrel if there is a thunderstorm or if it is very hot. Yours sincerely, Miss ...

The summer was passing. Just to get the message across generally, Sir John Anderson's office had issued an information sheet at the end of July 1939. Now starvation was the spectre, as well as gas attack from the air:

> YOUR FOOD IN WAR-TIME: You know that our country is dependent to a very large extent on supplies of food from overseas. More than 20 million tons are brought into our ports from all parts of the world in the course of a year.

In fact a lot of people did not know just how dependent. And much of that food from overseas was used to feed farm animals – for home-produced meat and milk. Pigs and poultry were also fed on cheap imported corn. The nations' pets came at the end of a long food queue. They were terribly vulnerable.

The first wave of refugee pets was coming out of quarantine. 'Marko', the St Bernard from Vienna, was reunited with his 'refugee mistress' at the Blue Cross Quarantine Kennels. She wrote gushingly: 'I have had the joy of meeting Marko again. Words are so poor to express what we feel in our innermost selves at such a moment. Hardly had I entered the Quarantine Kennels when I heard the barking of my dog. Out of thousands I should have recognised it again. It sounded so sad, and I was overcome to think that his poor dog's soul had ceased to hope for a reunion with me.'

Meanwhile Marko's 'knowledge of the English language is astonishing,' wrote his mistress. 'He understands each word much better than I.'

A trial blackout was held in British cities on 10 August. On the surface it was still jolly, super summer hols. At the London Zoo in Regent's Park there were elephant rides, buns and ice cream. The Children's Zoo teemed with kiddies. 'Twice a week the park was open late, you could dine elegantly and with courteous attention, then dance outside holding your partner deliciously close as you whirled under the coloured lights in the trees,' wrote one visitor. 'Half London had flocked to see the newly arrived giant pandas ['Ming' and 'Sung', supposedly a breeding pair captured in the wild in Sichuan by the American hunter Floyd Tangier Smith] and Princess Elizabeth and her sister Margaret made frequent visits.'

The news meanwhile was all about German demands on Poland and whether Britain and France really would resist them. In a gallant spirit of intervention, Lieutenant-

Colonel Gartside of the RSPCA rushed to Cracow in a bid to see what might be done for the Polish Army's many horses – 'but circumstances made practical action impossible,' he reported.

But the Colonel was able to make some fascinating observations about Poland's endangered pets. Because of the 20 zloty (16*s.*) licence, 'only prosperous classes kept dogs in towns'. But in the country, 'rural authorities charged nothing and every household has a dog and villages are overrun with mongrels. But they all seemed to be very happy, following workers in the fields or guarding flocks of geese.'

'Cats are not commonly kept as pets in Poland,' he reported. 'There are cats in many houses in the cities but they are not fed or looked after and are in a semi-wild state. They are seldom seen being very active and cunning, and live to a great age.' Polish horses, of which there were five million, were 'thin but hardy', he noted. They would get thinner.

There was a full meeting of NARPAC on 23 August (the day the Nazi-Soviet Pact was signed in Moscow), chaired by the affable Colonel Stordy. 'In view of the present emergency the chairman had called this meeting to discuss what immediate steps should be taken in connection with animals,' the police representative minuted. 'A scheme to deal with the practical consideration of animals stranded, injured or in the possession of children being evacuated should be formulated within a matter of hours. A full emergency scheme could be ready by the end of the week.'

Most concern thus far was with London pets. Stordy contacted old veterinary chums to be assured that Hull, Birmingham and Manchester were in the process of organizing something. The National Veterinary Medicine Association was co-operating fully.

It was agreed that a draft notice should be prepared for the press, the BBC and animal welfare societies. Headed, 'Measures to Meet an Immediate Emergency', it read as follows:

> If at all possible, send or take your household animals into the country in advance of an emergency.
>
> Should you decide to keep your animals with you, find out at once the nearest veterinary surgeon or local centre of an animal welfare society. The local police officer will tell you.
>
> If you have animals with you during a raid take them into the household shelter. Put dogs on a lead. Put cats in a basket or box.
>
> If you and your family have to leave home at very short notice (you will not be allowed to take animals with you under the official scheme), on no account leave them in the house or turn them into the street.
>
> If you cannot place them in the care of neighbours, it really is kindest to have them destroyed.

Officials from the newly established Ministry of Home Security approved the draft – 'The advice is in conformity with [our own] Handbook No. 12 and may be taken as sound.' On 25 August a certain Mr F. M. Hillier brightly noted that he had sent out a press notice – 'exactly conforming to the file note'. It was now headlined: 'Advice to Animal Owners'.

On that same day the agreement with Poland was signed offering British military aid in the event of an attack. It was meant to be a deterrent (German armies were massing on the border); it bought a couple of days.

This must be it, absolutely, irrevocably, or would there be another compromise? Evacuation of London's schoolchildren was imminent. Maybe pets would be made safe too. The existence of NARPAC and what was now called 'Official Advice to Animal Owners' was announced in virtually every national and local newspaper and on the BBC News the next morning. And there it was, that caring-seeming line about it really being 'kindest to have them destroyed'.

Amid grave announcements of military mobilization and diplomatic deadlines, it seemed a small piece of housekeeping, a way of doing one's bit. Put up the blackout curtains. Kill your domestic animals.

A national tragedy was in the making.

Chapter 3
Killed by Order

A nation of animal lovers stirred. The redoubtable Nina, Duchess of Hamilton, rushed from Scotland to London with a statement to be broadcast on the BBC. It went out on the morning of 28 August 1939: 'The Animal Defence Society is anxious to have the names and addresses of people who can offer free accommodation for other people's animals. There must be large numbers among those who are likely to leave London and provincial centres in an emergency who cannot afford to have their animals boarded out. It is emphasised that accommodation offered must be free of charge.' A statement appeared in *The Times* personal columns the next day:

> Homes in the country urgently required for those dogs and cats which must otherwise be left behind to starve to death or be shot. Remember that these are the pets of poor people who love them dearly and who will have sufficient worries without those caused by the knowledge that their pets are suffering.

'The result of the broadcast was startling,' Louise Lind-af-Hageby of the Animal Defence Society wrote: 'Within five minutes, the offer of a home was made by telephone and accepted.' But the offers were soon outnumbered by the animals which were brought to Animal Defence House [the Society's HQ in St James's Place, Mayfair] by their anxious and distressed owners – soldiers and sailors called up, families about to be evacuated:

> And so the house was filled day by day with ever increasing numbers of dogs and cats and other animals. There were monkeys, parrots and canaries. And owners did not only bring their animals to the Society. The private house of the Duchess of Hamilton was also besieged by people who clamoured for assistance in finding homes, who wanted to save their animals from air raids and destruction.

'If you can't take my poor Bob now, I must have him destroyed tomorrow morning,' said one. It was heartbreaking.

The Ministry of Home Security hated this freelance initiative. A NARPAC 'representative' evidently turned up at Animal Defence House the next day and insisted that all such appeals 'go out in its name'. The Duchess of Hamilton pointed out huffily that she had not even been asked to be represented on the Committee. Relations would not improve.

'Is there a pet in the house?' so the *Daily Mirror*'s Susan Day asked brightly on 28 August. She advised readers to 'make your plans for their comfort and safety now', but rather than follow the Government's soothingly lethal guidance, her humane feature article argued strongly against doing anything with immediate fatal intent. She recognized the acute dilemma faced by pet owners, asking perceptively: 'Is it that perhaps we feel a little guilty about our dumb friends that they should have to suffer for the horrors made by man?'

Yes, it was true that 'you cannot take them into public shelters,' she said. But there was plenty to think about first. 'If you live in an evacuable area, send them at once to friends who are outside the danger area,' she advised. 'The ODFL and NCDL have lists of people who are willing to accept pets and motor car owners who can assist.' If only it were that straightforward.

'Putting your pets to sleep is a very tragic decision. Only do not take it before it is absolutely necessary,' insisted Miss Day. 'If there should be no war you would feel terribly upset afterwards to think that you had parted with your little friend for no purpose.'

In London, newsmen watched the frantic comings and goings in Whitehall. A Chow dog was seen wandering unattended on the Duke of York Steps near the German Embassy. 'A passer-by gave him a facetious Heil Hitler! salute,' so it was reported. 'The dog wheezed and went to sleep.'

It was recognized as the pet of the recently departed ambassador, Joachim von Ribbentrop, who at one stage had had three Chows plus a Pembrokeshire Corgi (one Chow had been killed by a flying golf ball in Scotland in 1937). Ribbentrop was recalled to Berlin, leaving the remaining dogs in London, and was replaced by Herbert von Dirksen, the former ambassador to China, who was seemingly indifferent to pets.

That afternoon a 'mystery black cat'[3] was seen in Downing Street. 'When it reached the door of No. 10,

3 Secret Cabinet Office files reveal that two cats, 'Bob' and 'Heather', were employed at the nearby Treasury as mousers, who prowled the precincts of power as war clouds gathered.

onlookers cheered,' a newspaper reported. 'When photographers rushed forward, the cat fled.'

It was going to be war, surely, and better to follow Government advice. Many thousands of pet owners were doing just that. A nation of animal lovers stirred.

On Tuesday, 29 August, Mrs G. Blandford, a schoolmistress from Highbury, north London, noted in her diary: 'I seem to see such a lot of people taking cats about in baskets – evidently to be destroyed, as the people are obviously not going on a journey.'

A newly-wed couple in Kingston recalled the neighbours coming round and suggesting firmly that their dogs – an Alsatian and a Cocker Spaniel – be destroyed 'in case they ran amok in an air raid and bit them'. During this genteel-enough discussion, the dogs inconveniently did a lot of barking. Perhaps they sensed the gnawing tension building in the London suburbs. Their pets had to go; it was the kindest thing to do.

The next day at NARPAC's new Bloomsbury headquarters there was a frantic effort to get firearms licences distributed, organize cars, armbands, badges and identity cards. Actual anti-gas and rescue equipment was meant to come from local authority stores. No one seemed to know much about it but twenty-two designated veterinary posts – based in existing surgeries – were now on alert across London.

Evacuation of children from London was officially announced on the 31st – it would begin the next morning, 1 September, the day the German attack on Poland began. The British Armed Forces were mobilized.

'Whatever happens don't let us doggy people get the jitters,' Mrs Phyllis Dobson, editor of *The Dog World*, commented in the issue published that day. She complained meanwhile about a ghastly say-no-to-this-capitalist-war pamphlet she had just been handed on Westminster Bridge by, so she presumed, the 'Communists'.

'What [the possible outbreak of war] will mean to canine affairs is not yet known,' said *The Kennel Gazette*. While the Kennel Club journal observed: 'A German Dog Show is planned for October under the patronage of the Army High Command, although by then they may find they have other, more pressing, matters to attend to.' How prescient they were.

BBC Television at Alexandra Palace shut down at 12.35p.m. on Friday, 1 September 1939 (a Disney cartoon, 'Mickey's Gala Premier' was the last item to be screened). All that week the BBC had been making live outside broadcasts featuring Freddie Grisewood – from London Zoo at Regent's Park. The plans made at the time of Munich were being put in place, sandbags filled and war-veteran keepers, Overseer MacDonald and Keeper Austin, were reacquainted with the workings of the .303 Lee-Enfield rifle for fear of a great animal escape. Once again the heat in the reptile house was turned off. Sluggish snakes awaited their fate.

Newspapers carried stories of deep shelters for animals being dug, and plans for their wholesale evacuation to the Zoo's 'country estate' at Whipsnade in Bedfordshire, 35 miles north of the capital. 'Enormous crowds,' were reported, 'in spite of bad weather and inclement political conditions.' A name suggested for a Giraffe born on 28 August was 'Crisis'.

Thousands of children, evacuation labels round their necks, were saying goodbye to beloved pets. Blackout was declared across the country; the ARP was mobilized. How troublesome it all was for the judges and exhibitors struggling to get to the Harrogate Dog Show, highlight of the north of England dog lover's year. It went ahead nevertheless on 2 September.

There was, however, plenty to see in the gloom, including 'the Penyghent Alsatians in obedience tests', and a display by the sheepdog filmstar, 'Owd Bob'.[4]

As *The Dog World* noted: 'The Border Terrier bitch winning most Challenge Certificates in the breed just before Harrogate closed the shows was Mr and Mrs Jordan's champion Susan of Shotover, who luckily managed to get her well-deserved title before Hitler interfered.'

The whole thing was terribly inconvenient. Mrs Brinton Lee of London W10 had already foresightedly despatched her cat, 'Smout', by train to friends in the Cotswolds. On Saturday the 2nd, the Mass-Observation diarist left the capital by car for Oxfordshire with the kitten, 'Dibs', in a basket. Arriving at Moreton-in-the-Marsh, she discovered the village to be inhabited by strange, wide-eyed children.

'Our kitten soon disappeared,' she wrote, 'and we thought it had been picked up by the Liverpool evacuee children. We went out looking for it in the pouring rain while asking ourselves why should we be bothered [with] such a small thing as a kitten when war seemed so close.'

Actually, it would have been all that mattered. Then Dibs appeared, 'after dark, quite dry and cheerful'. The Merseyside urchins were blameless.

London was emptying, the children leaving by train, the better-off families, like Mrs Lee, scrambling by car for the country via 'the principal routes out of town [which] are

4 In fact Collie bitch 'Glyn', who starred in the hit film of the same name (*Owd Bob* in 1938) – a canine drama in which a sheepdog from Derbyshire and his owner (newly arrived in the Borders) must take on a grumpy Scottish farmer and his Alsatian, 'Black Wull', in the sheepdog trials. Glyn became hugely famous and made lots of personal appearances. His Alsatian co-star, 'Crumstone Storm', would go on to enjoy an even more dramatic wartime career.

one-way streets for three days', as an American eye-witness recorded. 'Cars poured out pretty steadily all day yesterday and today, packed with people, luggage and domestic pets,' she wrote on that same Saturday.

They were the lucky ones. In the Cotswolds, Dibs and Smout were already settling in nicely. There were plenty of domestic pets heading for a different destination.

*

It was a busy time too at London Zoo for all the wrong reasons. The Day Log for 1 September has 'Germany invades Poland' written in red ink. 'Animals unwell: Ruffed Lemur, Patas Monkey, Raccoon, Asiatic White Crane, Wolf, Dingo, 2 Elephants and Sun Bear, Eland, Camel, Llama, Reindeer.'

As the Zoo's director, Julian Huxley, wrote in his memoirs: 'When the news came over the radio the first thing I did was see that the poisonous snakes were killed, sad though it was for some snakes were very rare as well as beautiful. I closed the aquarium and had its tanks emptied and arranged that the elephants who might run amok if frightened be moved to Whipsnade.'

The Zoo's own ledger of 'occurrences' (written up each evening, an evocative record of births, deaths and transfers of its ever-changing population) for 1 September tells the sad tale of what happened. There were 593 visitors, of whom 68 paid 6*d.* extra to visit the aquarium. They were unaware of the slaughter in the reptile house – where 35 snakes were KBO'd (Killed by Order) – kraits, cobras, vipers, two puff adders, sundry rattlesnakes and five Gila monsters – all 'beheaded', according to one source.

That same day the pandas were crated up and began the ponderous journey by road to Whipsnade – with sundry chimps and Franz the orangutan. The next day Babar, the Asiatic elephant, would make the same journey north. The Occurrences Book also noted the arrival of a moholi galago (bushbaby), 'found in the neighbourhood and handed to the superintendent'.

It was reported on the 2nd: 'The poisonous snakes at the Zoological Gardens have been destroyed. The non-poisonous snakes were tended as usual but all will be destroyed at the outbreak of war.

'George the centenarian Alligator will be saved, along with the Chinese Alligator, the Komodo Dragon and the two largest pythons. The black widow spiders and the bird-eating spiders in the Insect House along with the scorpions have also been destroyed. Ming the young Giant Panda left for Whipsnade yesterday afternoon. Other animals will leave at the weekend.'

On Sunday the 3rd at 11.15 a.m., Neville Chamberlain announced on the BBC that Britain was at war with Germany. One London woman would later claim her dog got out of its basket and 'stood to attention at the historic moment'.

That same day Winston Churchill was remade First Lord the Admiralty, political head of the Royal Navy. In the venerable building at the Trafalgar Square end of the Mall, a nameless battleship-grey cat prowled the basement. He would soon make Mr Churchill's acquaintance. Their relationship was to be full of incident.

The Zoo's Occurrences Book was written up: 'Gardens closed at 11.00am on declaration of war against Germany. Remaining closed until further notice.' That same day the non-poisonous snakes, pythons and anacondas – huge creatures – went the way of their venomous brethren as the sirens howled. The aquariums were closed ('danger of flying glass') with the fish released or destroyed. Some reportedly went to West End restaurants.

Fear of interrupted food supplies seems to have determined who was next on the Regent's Park death list. The manatee in the tropical hall of the aquarium got it, as did six Indian fruit bats, seven Nile crocodiles, a Reeve's muntjac and two American alligators ('destroyed owing to war conditions'). Two lion cubs were put down.

Meanwhile over at the Kursaal Zoo at the Southend funfair, Mr Frank Bostock, the attraction's owner, complained that the animals, including seven lions, bears, wolves and a tiger worth £1,000 were now, 'practically valueless and a liability'. An RSPCA Inspector, Mr T. Stephens, despatched them all with a rifle. The fate of London's big cats looked distinctly grim.

Regent's Park continued as it might without visitors except for a few curious journalists. The animals were 'bored and disgusted, listening and waiting for the crowds that feed them,' so it was reported. They were getting 'more and more depressed'. The appearance of barrage balloons had vexed them more – 'although some of the big cats in the outdoor pens interpreted them as some sort of potential prey'. The exodus to Whipsnade continued, while common native bird species, kites, kestrels and herons were simply released into the wilds of north London.

According to the daily log, an alligator was sold for £6 to 'Koringa', the notorious female fakir of Bertram Mills Circus. At Whipsnade the elephant, 'Jumbo II', was shot 'because of housing difficulties,' according to the *Daily Telegraph*. It was healthy young bull, 'a gift from the Governor of the Soudan'. Presumably the influx from London meant the prospect of a noisy territorial dispute. On the 15th Regent's Park reopened to the public but few people visited.

The public did not come because the capital's children had been evacuated, so the naturalist James Fisher explained in an article in *Animal and Zoo News*: 'The people left in London have little time for anything other than work and sleep.' So why not shut the Zoo down?

Not everyone was a zoo lover. An anonymous columnist for *The Animals' Defender*, 'Anti-Vivisectionist', went to see for himself. After he 'had paid sixpence and passed the turnstile' he encountered, 'a creature coated with short, bristly fur, a Siberian badger, according to the label.'

> Up and down, up and down the front of the cage he went, pausing momentarily every few seconds to dart his snout through the impassable metal meshes. Some of the visitors offered him food but it was not food he wanted. I saw nothing but dreary, weary creatures destined to live, and then at length mercifully to die, deprived of everything that could give their life a meaning.

'It is not only as the proprietor of a menagerie that the London Zoo carries its weight in organised science,' wrote James Fisher. 'Some people might think it wasteful and wrong that academic science should still be indulged in when Britain is fighting for its life. But honest free knowledge is part of what we are fighting for.'

The country retreat at Whipsnade would prove the London beasts' lifeline. The big cats and bears would be spared. Other animals would be less fortunate.

Chapter 4
Killed by Kindness

West Ham, the County Borough on the eastern edge of London, was, on the eve of war, a tangle of industry and housing along the Rivers Lea and Thames. It contained the Royal Docks and sprawling factories. It was where, over the years, the malodorous activities of London had been exiled: slaughterhouses, fat renderers, glue and fertilizer works.

It had waxed fat on the proceeds, raised a splendid town hall and laid out its own electric tramways. Its fleet of municipal horse-drawn rubbish carts had been partially mechanized in 1936 to tend the streets of humble terraces but still the borough employed hundreds of horses.

Messrs Price's Bakeries had eighty horses, United Dairies had ninety in Freemasons Road and the Co-op fifty horses in their Maryland Street stables, dutifully delivering the bread and milk. There were dairy farms in the borough, piggeries, cattle yards and abattoirs, plus a large number of household animals, fowls, rabbits, etc. Every household was reckoned to have a cat, 50,000 of them, with 10,000 more in warehouses and factories.

In 1939 West Ham was of distinct interest to *Luftwaffe* strategic planners because of all those docks and chemical works. It was an industrial target. And like the whole of the East End of London, it was full of animals.

In the archives of the RSPCA is an unpublished typescript history of animals under fire in the borough, unsigned, but almost certainly by Harold Edwin Bywater, chief veterinary officer of West Ham and NARPAC treasurer. He was a proper cockney, the son of an East End haulage contractor who had worked with horses (for which he clearly cared deeply) as a teenage Army Veterinary Corps 'shoesmith' in the First World War and then qualified as a vet. Intensely personal and humane, it begins with the first intimation that they were both a target for bombing and had a special problem with animals. Thus in early 1939 a volunteer animal ARP was organized, recruited largely from butchers and slaughtermen. A special plan was made for Silvertown on its dockland peninsula, which could easily be cut off by bombing. A mayoral fund paid for humane killers but not much else. First aid classes were given at the town hall. The vicar of St Marks in Silvertown, the Reverend Shaw, gave over the church hall to pioneering animal ARP lectures.

The borough was also home to several firms, Harrison, Barber & Co. and the Merton Bone Co., who been turning the carcasses of animals into fertilizer for over half a century. Messrs W. J. Curley of Marshgate Lane had fifty men rendering animal fat into soap beside the River Lea. They might cope with 100 tons per week. The typescript came to what happened next:

> On Sunday 3 September, at 10 o'clock in the morning, a final briefing was held at the Town Hall for the principal workers in the animal service. The meeting had just terminated when the sirens sounded heralding the commencement of war.

> Almost immediately the town hall became besieged by a crown of panic-stricken people bringing their animals for destruction.
>
> In spite of attempts to reason with the well-nigh hysterical mob, we were soon inundated with dogs and cats whose owners left them in the corridors and offices. Which was the commencement of several days spent almost entirely in the destruction of abandoned animals.
>
> At six o'clock in the evening, a message was received with effect that Harrison, Barber & Co's premises were filled with the carcasses of dogs and cats, and vehicles were loaded with more carcasses lined up outside their factory. The yards were covered to the depth of several feet with the bodies of dogs and cats. The weather was warm and sultry and the odour was perceptible over a wide area.

It had all been done with good intentions, the press release, the BBC announcement – 'it's the kindest thing to do' – the stockpiling of humane killers, but it was not meant to be like this.

A despairing Mr C. H. Gaunt, the Superintendent at Ilford, told a reporter that they were doing 'the best thing for animals by destroying the pets of evacuated families.' PDSA staff were 'working 18 hours a day'. They had destroyed 300,000 London animals in under 10 days. Of its reluctant role the charity says 75 years later: 'PDSA performed a public service that many would have shied away from. Pet owners looked for support and PDSA was there.'

Mr Bywater knew who it was to blame. 'A few days before the BBC advised the general public to have their cats and dogs destroyed and as a result the situation became chaotic,' he wrote. 'When officials from NARPAC and the Ministry of Home Security went to the factory on Monday [4 September] the conditions were so disagreeable they found them impossible to bear,' he continued. Bodies were sent by lorry to other places with disposal facilities – Fulham and Brighton – until they too were overwhelmed. They would have to be buried in pits – but where?

That day Colonel Stordy told the Ministry that 'over two and a half million animals had been slaughtered and transport arrangements had collapsed'. NARPAC struggled to stop the slaughter with a press release on the 4th: 'The committee urge owners of animals not to arrange needlessly for the immediate destruction of dogs and cats. An animal is quite safe in a gas-proof room, and its keen sense of danger may be of help to people in the room.'

But one London dog got away with his life. When the German Embassy was evacuated on 2–3 September, the embassy dog 'Bärchen' ('little bear') got left behind. It was the dog pressmen had seen being given 'Heil Hitler!' salutes the week before. 'When the embassy closed nobody noticed, it dragged around matted and dirty'.

Policemen reportedly fed poor Bärchen on 'ham sandwiches and bones brought from home' as he wandered in the street where his plight was picked up in a newsreel. When it was reported that he was to be destroyed by the RSPCA, 200 people asked to have him.

Dennis Mulqueen, an Irish-born second footman from the embassy, swooped. A little later he was reported to have taken him to a new home in the suburbs, where he was 'now recovering from eczema caused by eating too much meat'. The *Daily Mirror* gave him a newspaper column to comment on German affairs but, as his further adventures were limited, this did not go far.

Coming in the opposite direction was Sir Nevile Henderson, the pro-appeasement British Ambassador to

Berlin, along with 'Hippy'[5], his Dachshund. Hippy went into quarantine. *The Times* at the same time published an appeal from Mr D. L. Murray of Arundel Terrace, Brighton, that his Dachshunds had been 'the subjects of insulting remarks' on the streets of the seaside resort. The RSPCA pleaded for common sense, suggesting anyway that the breed originated in France.

In London the slaughter of the innocents continued. Manchester caught the frenzy. Those faceless officials who had so casually triggered it tried what they might in further press briefings to shut it down. *The Times* commented on the 7th: 'A widespread and persistent rumour that it is now compulsory to get rid of domestic animals is causing many thousands to be taken for destruction. Centres run by animal welfare societies are filled with the bodies of animals. The National A.R.P. Animals Committee emphasize that there is no truth whatever in this rumour.

'Apart from everything else, the huge destruction of cats that is continuing at present may lead to a very serious and increase of vermin, such as occurred in Madrid and other Spanish cities.'[6]

Stordy reported to the Minister on the 11th: 'All estimates were overtaken by the wave of destruction that took place.

5 When Hippy died not long afterwards, Sir Nevile wrote: 'I can hardly conceive of another life unless Hippy be waiting there to share it with me.' A biography of the Dachshund would be published in 1942, including a claim that he had growled at Nazis on the streets of pre-war Berlin.

6 There was a story that a 'shock force' of stray cats had been collected from all over Spain to deal with a plague of rats in the city, besieged from October 1936, which eventually fell to the Nationalists on 28 March 1939. A small PDSA team, driven by an American war correspondent, had got into Madrid just before the end. They found plenty of wounded mules but no cats or dogs, 'which bore out stories that they had all been eaten,' said their report. 'Mules which had been humanely destroyed quickly disappeared for their flesh.'

Incinerators were not enough. Many thousands of carcasses were dumped on the ground on land[7] being reclaimed at Ilford [next to the] PDSA Sanatorium under the control of Mr Gaunt.' He estimated at least '80,000 in one night'. The burial of the carcasses had been 'very efficient'.

'With the advent of war, it was not only the poor that took advantage of the clinics,' wrote Stordy, 'but many of the well-to-do have had also their animals destroyed' (he pencilled '¾ of a million' in the margin).

The destruction was London-wide. Society pets as well as strays went into the lethal chamber. The RSPCA reported: 'From 1 September all the Society's clinics were working day and night. A temporary euthanasia centre was even opened at Headquarters [105 Jermyn Street in the heart of the West End].' Mary Golightly, *The Dog World* columnist, reported:

> The stories I am hearing seem almost incredible. One veterinary surgeon at length refused to put any more dogs away. The dead bodies were stacked in a heap outside his surgery waiting to be shovelled into a van and taken to the council incinerator.
>
> They were not all mongrels (not that that matters at all) but there were beautiful, highly bred specimens among them. Dogs of the show bench, dogs that had been treated with the greatest care, pampered, hair brushed every day in place, and now this!

In *Woman* magazine, the popular novelist, Christine Jope-Slade, parlayed the pet slaughter into a romantic short

7 Stordy would tell a newspaper, 'It was the greatest single burial of pets anyone has ever seen. The secret burial ground was just by the site of recent boring for an Underground railway extension so we were able to run trucks down there.'

story called 'Enemy Alien'. A handsome young veterinary surgeon, 'Charles Maurice', calls on the heroine, 'Mollie Dresden' of Redmayne Kennels, to put down her Irish Terriers. 'I must keep the youngest and the best and hope to breed from them after the war,' she tells him. 'Two of the beagling packs have gone. I had the Master over here yesterday.'

Every vet's surgery, every animal clinic was besieged. The PDSA reported: 'Long queues lined up at [our] dispensaries. People said that if we would not destroy them they would turn them loose into the streets.' One man brought a pair of Dalmatians and when the clinic refused to kill them, he left them 'tied to the railings of the police station with a note authorising their destruction'.

The Canine Defence League starkly called it the 'September Holocaust'. 'Looking back on those dark days our men still shudder,' reported its journal, *The Dogs Bulletin*, 'The clinics have always been centres of healing, now they were being turned into centres of destruction.'

Phyllis Brooks was the young wife of a Dumb Friends' League ambulance driver. She later recalled: 'War came on a Sunday morning, and I went with him [to the Wandsworth shelter at 82 Garratt Lane]. The sirens had gone, which had upset people. There was already a queue of at least 50 people with their animals, as well as protesters trying to dissuade them from having their animals put down. Many were broken hearted about it.'

A 24-year-old civil servant living in Croydon took a 'kitten that had been hanging around for several weeks, crouching in our shed' to a Canine Defence League clinic to be destroyed. The family cat, 'Tiger', was enough to worry about in uncertain times. She recorded the bleak encounter for Mass-Observation:

> An elderly man was in charge and I had hardly begun to explain before he ushered us into a small room where he apparently put the cats into the lethal chamber straightaway.
>
> He then said his first words since I had been there. 'Name and address please.' I gave him the particulars and he put them in a book beside a long list of other names, nearly all with the word 'cat' against them.
>
> I gave him a shilling and he entered it in a book. I went away feeling very sad. He had a very miserable expression about him as if he was absolutely fed up with the number of animals that were being brought in to be destroyed.

Major Hamilton Kirk, a prominent north London vet, would tell *Dogs World* of this, 'the most dreadful and loathsome experience' in the whole of his professional career. It was not just dogs and cats that were going up in smoke. 'Only this week I have had the misfortune to have been called to destroy two lions and five monkeys,' he wrote in October. 'In the case of a third lion, it was the tamest I have ever seen. As I approached the cage, it came out of its inner house with its meat tin in its mouth, played about with it, rolled on its back and purred.

'How could one, in cold blood, take the life of such an animal?' Killing a monkey was to feel something like murder,' he added. 'Think of the feelings of a vet when asked to destroy a happy little dog which jumps up, wags its tail and licks one hand. It is a dreadful business.'

Dr Margaret Young, leading spirit of the Wood Green Animal Shelter in north London, recorded in the first days of war – 'a queue nearly half a mile long of people who had to part with their pets'.

There were plenty more pets locked in houses by

fleeing families or simply abandoned in the street. Dr Young appealed for funds to help the 'scores of animals left behind and slowly starving to death. We know of cases of cats shooting up women's shopping baskets in a vain endeavour to find food. And there are dogs, little more than skeletons, hunting the dustbins, hoping to find some scraps.'

The London Institution for Lost and Starving Cats based in Camden Town reported: 'Staff pleaded with owners not to have their animals destroyed but they were adamant. Such people were kindlier perhaps than many others, because staff have continually been called to houses which have been evacuated to rescue some wild, starving cat who has been left behind.'

Abandoned cats would haunt the capital for weeks to come. Our Dumb Friends' League alerted newspaper readers in October to the plight of 'imprisoned cats' still shut up in houses, and appealed 'to owners who have inadvertently left their cats behind, or to people who know of such cases, to write to the League'. No names would be mentioned, no prosecutions brought.

The first of many wartime cats-being-turned-into-fur-coats rumours took flight. And there were more. A cleaning lady in Hampstead was reported as saying:

> You know what they're doing with all them cats that's vanished? They're using the skins to make British Warms [military reefer coats] and they boil down the fat for margarine. They say there's cat in pies.

There were glimpses of kindness amid the carnage and abandonment. The RSPCA swooped to rescue the pets of London County Council schools (which were still on summer holiday). 'Over 500 school animals including an alligator which was referred to a zoo, were evacuated to the Horses' Home of Rest at Boreham Wood and to The Ember Farm, Thames Ditton,' reported the Society. 'Every animal was carefully labelled.'

Less fortunate perhaps were the 'experimental animals' at The London School of Hygiene and Tropical Medicine – 'eighty cats, monkeys, rabbits and other animals [which] were duly taken and humanely destroyed' by Our Dumb Friends' League.

The *Veterinary Record* gave advice to practitioners on how to dispel fears of those clamouring for the destruction of their pets: 'The sound of gunfire was very similar to that of thunder, which they knew about anyway,' and 'there are distribution and evacuation schemes like that promoted by the Duchess of Hamilton.'

There was indeed an alternative for the astonishing Duchess had waded in again. The Oskar Schindler of pets had made her dramatic appeal on the BBC on 28 August. Now, as would be written by her lifelong collaborator, Louise Lind-af-Hageby:

> Animal Defence House [the Society's Mayfair HQ] was filled day by day with ever increasing numbers of dogs and cats and other animals. There were monkeys, parrots and canaries. The door-bell and the telephone rang ceaselessly. A procession of applicants waited for the opening of the offices.

The Duchess opened her own substantial London home, Lynsted in St Edmund's Terrace, just north of Regent's Park (Louise Lind-af-Hageby lived next door at No. 8), as a clearing station. Mary Golightly was a volunteer. She recalled the perils of the first week of war when, 'we had to collect the dogs from different parts of London and drive them to Regent's Park, the whole time getting mixed up with columns of evacuation school children and then getting smacked in the eye by the continuous one-way streets.'

'These are brave people,' she wrote, 'these people I took dogs from in Stoke Newington, Poplar, Acton, Islington, all giving away a part of their very hearts as they placed their pets in my arms or urged their big fellows to get into the car.' They called the Duchess, 'the lady for the dogs'.

She described the scenes at Lynsted, where multiple dogs, 'lie quietly tied up in various points in the London house of the Duchess'. There was also a cattery extemporized in the shrubbery.

'You would be amazed to see them,' she wrote. 'The Duchess, her eyes full of tears, places a tender hand on each and every head, moves from one to another with pans of water, bits of biscuit.' Her kindness had: 'Moved many a soldier. "Bless that lady of the dogs," said one. "It puts a bit of cheer into a feller when he knows his missus and his dog is alright".'

The Duchess's original scheme was for a diaspora of dogs (and cats), a call on rural pet lovers' goodwill to take in urban refugees. This was still the case. 'Letters offering homes in the country came at the rate of 200 a day,' she wrote. 'Letters begging the Society to take charge of animals came at a much greater rate.'

There was a second BBC broadcast, 'an appeal to owners of private cars to help with free transport. This brought much generous personal service and help.' Commercial 'kennels and cat-boarding establishments in the Home Counties' were also engaged.

It was by no means for posh pets only. This was a genuine philanthropic intervention to help the animals of the poor. As Mary Golightly wrote: 'It took a great deal of tact to persuade some who brought their dogs that our work was purely for those owners who could not afford boarding fees, but everyone was very nice.' One feels sure that they were.

Then there was the barking. In an open letter of apology to her St John's Wood neighbours Her Grace wrote: 'I would like you to know that I heard officially that last Sunday [3 September] thousands of dogs and cats were destroyed and two days ago three truck loads of dead bodies went out from a certain animal clinic in London. We should be horrified if this sort of thing happened abroad. How we can explain such a thing to our foreign friends in this so called animal-loving England?' (The journal of the Reich Animal Protection League would report in spring 1940 that 'millions of dogs and cats were killed in the first weeks of war … 750 tons of carcasses had been turned by one London firm into manure,' but did so in shock and sadness rather than propaganda gloating. It had been 'totally unnecessary'.)

'Considering how all these dogs are strangers to each other, and to us, I think they are all wonderfully quiet,' she wrote. Her neighbours might have disagreed.

Beyond Lynsted lay Ferne, the ducal estate in Wiltshire acquired in the 1920s. Nina Hamilton declared it once more to be a sanctuary for pets, as in the Munich crisis. The first raucous wartime transport of dogs and cats left London in the Duke's Daimler on 4 September. Many more would follow.

Chapter 5
Keep Calm ...

After a week, the killing frenzy burned out. Pet lovers blinked. What have we done? The sky had not darkened with bombers. A kind of calm returned. Where now were faithful Bonzo and little Oo-Oo? Reduced to fertilizer in Sugar House Lane or interred in some cold field in Ilford.

In Memoriam notices appeared in the pet press, looking it might seem, for forgiveness. 'In ever loving memory of our dear Chum, put to sleep to save from suffering 25 August, 1939,' said one. 'In memory of Bobbie. Put prematurely to sleep 30 August, 1939,' was another.

There was a meeting of the NARPAC executive on 12 September. Everyone threatened to resign but not over the massacre. Major E. J. Stuart, the Committee's transport officer, declared he could not work with Colonel Stordy. Mr Bridges Webb and Keith Robinson threatened to withdraw the PDSA and Dumb Friends' League outright – because they could not allow their 'clinics to be given instructions except during the actual course of an air-raid'.

Colonel Stordy insisted that unless he could give on the spot directions that would be complied with, he would find it impossible to continue, and tendered his resignation. It was not accepted. The Dumb Friends' League's inspectors, the Home Office noted meanwhile, 'had in a short space of time acquired an extraordinary reputation for a complete lack of tact or discretion'.

Some kind of order returned. The Metropolitan Police official on the Committee drily noted: 'It is something that they [the animal charities] have survived the outbreak of European hostilities at all.'

Was it even worth continuing? 'It would be advantageous in many ways if we were to be out of their disputes,' the Scotland Yard man noted, '[however] it would seem essential that we are continued to be represented on this committee.'

Money was the immediate problem. The Home Office would advance a mere £1,000. The Dumb Friends' and the PDSA proclaimed there was no need for public money as the charitable public had given enough already. They confessed to being replete with funds. A dedicated bank account would be opened with an overdraft facility, which the ODFL and PDSA would guarantee. The RPSCA and National Canine Defence League refused, as did the Dogs' Home, Battersea because 'we can only use our funds for the purposes for which they were given to us'. Charity began at home.

But they had to have some future source of income. Edward Bridges Webb of the PDSA had an answer. The new organization would function like any other charity – with a money-raising Appeals Committee. He came up with a splendid wheeze, 'registration' of the nation's domestic animals in a giant central directory with identity discs, distributed by local unpaid helpers, the 'animal wardens' idea, in return for voluntary donations. He calculated it would raise £10,000 in London alone. Whoever controlled the registry would control the nation's pets.

There was already something like it, the Tail-Waggers

Club, a populist, dog-enthusiast organization founded in 1926, partly sponsored by Spratt's Pet Products, which issued a name-engraved collar badge of its own bearing the Club motto 'I Help My Pals'. It published an engaging magazine featuring articles, some of which were ostensibly written by pets. Its headquarters were in Barking. But although the royal Corgis were members, it did not seem quite serious enough for the Ministry of Home Security. Actually, as it would turn out, the Tail-Waggers might have done better than anyone else.

The civil servants were delighted with the registration notion. It could all be paid for out of the British love of pets and there was no denying the strength of that particular passion. Even vets would work for free. It was noted for the minister that the People's Dispensary had 'always quarrelled' with the veterinary profession – but last month an agreement had been reached (a drawn-out legal action for slander had been settled and the Dispensary agreed to call their unqualified staff 'Technical Officers').

So let the Appeals Committee have what they wanted. They should be allowed to do door-to-door collections. NARPAC indeed should be the sole animal welfare fund-raiser. The RSPCA was incandescent.

The message meanwhile that killing pets was wrong was at last starting to work. Pet lovers faced up to meeting the challenges ahead with their animals by their sides. Each did so in his own way. 'Do nothing in a panic! We urge everyone not to destroy their studs, whether of rabbits, cavies, mice or cats,' editorialized *Fur and Feather* magazine on 8 September. 'Nobody knows how long the present emergency is going to last. We must have more rabbit breeders. Carry on!'

The magazine's exhortations were for readers to breed their pets so as to eat them. 'Every breeder in the country should now be making plans to produce rabbit flesh,' said the journal. 'No matter whether his stud consists of purely fancy varieties, of fur, or of wool rabbits, he can utilise a part of it for food production.'

'To keep rabbits is to perform a national service,' said the editor of *The Smallholder*. 'Soldiers are we now, every man and woman amongst us.'

Mr C. H. Johnson, president of the National Mouse Club, declared: 'This is not just a nod to those fanciers who have answered the national call and joined the services but also to those of our members who are left at home. I ask you to continue with mouse activities. Let us make a solemn resolve that we will always keep a few mice, however difficult it may be.' And who could argue with that?

Cats too dug in for a long campaign. Captain W. H. Powell, the writer of 'Cats and Catdom' (the feline spot in *Fur and Feather*) declared on 15 September: 'We MUST strive to keep the Cat Fancy going through this infernal business. The cat fancy is not some useless luxury hobby. Do nothing irrevocable!'

The Cats Protection League announced stirringly: 'There must be no truce in the war to help cats.' Mr Albert A. Steward, writing in its fine journal, *The Cat*, recognized that it was 'difficult in the present tragic days to write about the ordinary lives and needs of cats. The human tragedies, mental and physical, that are about to surround us will be uppermost in all minds and the little companions of our peaceful days will be forgotten by many.'

It was not the Nazis who cats should fear but 'indifferent, bad and nervy owners,' said Mr Steward. He defined as such 'those who, when rationing comes, will make no effort to feed their cats' or 'those who, when air raids are expected, will be too lazy to ensure that they are brought in at night'. Other owners 'will be so scared that they will forget everything but their own fear,' he predicted. The cats of such uncaring owners would, 'find life unendurable and wander away'.

It was not potential enemy action immediately endangering cats but changes in the ordinary routine for such a creature of habit. Evacuation and blackout were disturbing enough for humans. 'The blackout has affected the town cat more than his country cousin,' *The Cat* would record, 'who, indeed, generally gets the best of it, always.' But there was a distinct class order for wartime felines:

> If the plebeian city puss, whose playground is the street, is not in by blackout time he must stay out, and there we must leave him. Next in the social scale is the flat cat. He has had to forgo the pleasure of taking the air on the balcony or window ledge at dusk, a time of special interest to ancestrally nocturnal animals.
>
> The patrician cat, who lives in a house with a garden, has hardly felt the blackout after the first few weeks. The country cat has been affected by the blackout as little as the town aristocrat, except that motor cars in country lanes have taken a greater toll of life.

As for evacuated cats: 'Those who accompanied their owners into exile have only experienced that as might happen to them at any time,' said *The Cat*. 'The difficulty of settling in strange surroundings can all be surmounted by common sense – paw-buttering, extra fuss, familiar cushions or baskets and all the usual devices.'

Cage birds too might find safety in the country. *ARP Journal* declared at the end of the year: 'There is a list of animal lovers who will take birds or beasts in reception areas. Bird fanciers are opening their aviaries to town budgies and canaries and these kindly folk charge 2d per week for seed.'

Dogs would have to tough it out wherever they were, those that had survived the September massacre at least. In the first few days, pedigree and mongrel alike had gone to the lethal chamber. There followed a deeply unpleasant interlude when the survival of the poshest seemed paramount. Lesser breeds had better watch out.

A 'world-renowned authority and judge' wrote in *The Dog World*: 'I should say that the present time offers an opportunity to wipe out all mongrels and cross-bred dogs, and if the authorities could or would carry that out, it would clear the streets of a lot of danger and filth as well.'

'I do not want mongrels to multiply because they are ugly, ill-mannered curs, usually dirty and cross and have no value whatever, either to the senses or the pocket,' wrote the uncharitable dog expert, Mr George Wallwork. But those Nazi-seeming sentiments were soon stifled.

The Dog World canvassed pedigree breeders for their views in a round-up called 'Dogdom and the War' on 22 September. They were a little kinder.

Mrs J Campbell-Inglis, of the Mannerhead Poodles, Wimbledon Common, said: 'I suppose I shall carry on but I find a lot of people are tired of dog shows, which cost a lot and are not always much fun.'

Mrs E. M. Buckley of the Adel Chow Chows, Stratford-upon-Avon, said: 'England breeds some of the best dogs in the world and we'll need them when all this wretched business is over. By all means the weaklings should go to leave room for the best ones of the future.' An unpleasant sentiment Herr Hitler himself might have seconded.

On 29 September Mrs I. M. de Pledge of the Caversham Pekingese declared: 'I am determined to hang on to my kennel of Pekingese at all costs, especially my best studs and bitches. I have an Anglo-Nubian goat, whose milk will be invaluable for mixing with their biscuits, rice and meat.'

Mrs Ethel E. Smith of the Leodride Bulldogs said: 'It is early days and it might not be wise to embark on a plan which might cause endless regrets if the war proved to be a short one.' Meanwhile: 'For feeding purposes, Mrs Smith recommends good raw lean beef, occasional eggs and a good biscuit meal.' Such luxury was not going to last long.

The Duchess of Newcastle told *The Dog World* that she had twelve evacuated boys staying with her, 'luckily a very nice lot'. Her Grace was 'keeping on as many Clumber Spaniels and Smooth Haired Fox Terriers as possible but is worried that as time goes on, both Smith, her kennel man, and her chauffeur will be called up. But she is determined to carry on by working herself with perhaps the help of a kennel-maid.'

Chapter 6
... And Carry a White Pekingese

For an evacuated pet in 1939 you could not do much better than end up at the Duchess of Hamilton's Wiltshire animal sanctuary. After the first frantic weeks of September the place was teeming. At least there was plenty of room. 'A couple of hundred dogs are housed in the enormous coach house,' wrote Mary Golightly. 'Two hundred cats were housed in the private aerodrome,' so it would be reported, 'each one of which was as carefully looked after as the dogs [and] evacuee parrots.'

The 'aerodrome' at Ferne had been cleared by Lord Clydesdale, heir to the dukedom and intrepid pre-war aviator. A similar field was created at Dungavel House, the ducal home in south Lanarkshire (a former hunting lodge, adopted when stately Hamilton Palace was demolished in 1919, and definitely no hunting there now), to keep the family in airborne contact. This would prove significant in the course of the wider war.

'The secretary at Ferne is Miss Judy Mussprat-Williams, the fiancée of a flying officer somewhere in France, who is delighted with the more work she has to do as it takes her mind off her own affairs,' so Miss Golightly reported. Special commendation was also made to 'Miss Bunbury' and 'Miss Jacobs', pet lovers who took in evacuee dogs on a slightly less ducal scale. And Mr Bernard Woolley, impresario and dog-loving owner of the Lido Cinema, Bolton, had offered homes to 100 dogs. Apparently they were driven north in yapping car-loads by the 'head of cats', Miss Molly Atherton, 'in batches of 10 to 20, unloaded, given a run and taken into the cinema'. I wonder what was showing?[8] Actually they were promptly introduced to prospective hosts and taken into the bosom of Lancashire dog-loving homes until the shortage of

8 The evacuees would have plenty of choice because 1939 was a big year for dog films. They might have enjoyed *Peace on Earth*, an MGM cartoon parable released that year, in which the human race extinguishes itself in war and animals take over. Of UK productions, *Owd Bob* was still going the rounds, joined in July by *Border Collie*, a Cumberland-set semi-documentary, narrated by its hero 'Jeff' and followed promptly by *Sheep Dog*, which took the formula to Wales and featured 'Mr Tom Jones' the shepherd, his horse 'Tufty' and dogs 'Scott', 'Guide' and 'Chip'.

The big news from Hollywood was *The Wizard of Oz*, in which Dorothy Gale's dog, 'Toto', was played by a brindle Cairn Terrier bitch (real name 'Terry'). Alsatian star 'Rin Tin Tin (III)' had outings that year in *Law of the Wolf* and *Fangs of the Wild*. Yorkshire author Eric Knight meanwhile was in the middle of writing the short story, 'Lassie Come Home', that would be filmed in 1943 to establish the most famous dog movie franchise of all.

More urbane dogs in the Bolton audience would have enjoyed Myrna Loy and William Powell in the 1939 release, *Another Thin Man*, featuring the famous Wire Haired Fox Terrier 'Skippy' as their pet 'Asta'. Already his appearances in the series and in other films had created a huge interest in the breed in both the US and UK.

My own 1939 favourite would have been the peerless *Society Dog Show*, in which Mickey Mouse enters Pluto. While Mickey grooms his mutt, Pluto starts swooning over 'Fifi' the Peke. Things get worse before they get better. In contrast, pre-war feline movie stars were thin on the ground.

petrol made the flight from London much more difficult. Kind Mr Woolley!

The Duchess described her own charges: 'The evacuated cats were very numerous. Sometimes they would arrive with a dog friend.' Mrs Freeman's six cats, evacuee-veterans of Munich, were back, along with a long procession of 'grey pussies, black pussies, tabby pussies, white pussies, orange pussies, tortoiseshell pussies, long-haired Persians and shorthaired cats'.

'It became a very real problem as to how to distinguish each one,' she wrote, 'for cats have ways of slipping off collars. Many kind hosts came forward, but the greatest number of cats went to Ferne Sanctuary.' That winter there was an outbreak of feline influenza brought in by an 'evacuee kitten' – but the usual eighty per cent fatality rate was brought down to ten by 'the dedication of Miss Dukie and Miss Swallow who nursed the invalids day and night.'

There was embarrassment when the Duchess was fined £10 for blackout offences. It was reported that she had '200 evacuee dogs at Ferne House and has twenty girls, acting as kennel maids'. Twenty windows were recorded as showing lights over a number of nights. Her solicitor said, 'It is extremely difficult to control these strange young ladies.' The chairman of the bench, sentencing, said: 'It is a serious case. The fact that the Duchess has taken on a large number of dogs is not much of an excuse.'

Blackout was more of an issue for town pets. Motor vehicles, even in their reduced numbers, were a mortal danger. And dogs must be 'exercised' in darkened streets. This could be vexatious in several ways. There were many slithering upsets as the nights lengthened.

The Dog World reported at the end of September a new interest in 'white dogs – such as Sealyhams and Bull Terriers, useful in avoiding pedestrian collisions'. 'Carry a white Pekingese,' an opportunist Peke breeder advertised. The Dumb Friends' patron, Lady Hannon, devised a natty white saddle cloth for dogs while the *Daily Mail* promoted 'a white coat for your dog to protect him from unseen feet on crowded pavements in a blackout'. Another such garment came with jingling bells.

The National Canine Defence League offered the 'Lustre Lead' – 'glows with a beautiful fluorescent green colour in the dark – nothing to rub off or harm the dog – obviates risk of accident'. In the constant collisions and trampling dogs on leads proved as much of a menace in the gloom as those without them.

'Humans must learn to be cats and walk in the dark,' wrote 'Lucio', the *Manchester Guardian* columnist, a theme cheerfully picked up by *The Cat*. The editor pointed out:

> We should all know that while cats can see in dimmer light than we can, they cannot see in total darkness. I was troubled about my cat, a strong-sighted animal, when the lighting regulations came in. Not only was he nervous and worried about the pitch darkness, but he was quite unsighted and fumbled his way through the house.
>
> He specially disliked having to jump into a dark room from the window and I struggled with a torch and the black-out curtains in an attempt to help him without breaking the regulations. I then hit upon the simple device of a nightlight on the floor, where it cannot show outside. I commend it.

Cats and dogs could adapt to the blackout – but what about rationing? It would be food, not enemy bombs, that would determine the fate of pets in the much more testing times to come.

In those innocent, early months of conflict, 'feeding dogs in wartime is no problem,' the Bob Martin Company could say in a useful pamphlet. 'Several breeders during the last war kept their kennels healthy on a diet which consisted mainly of potato peelings and meat offal. We are officially informed by the Ministry of Food that there is no present intention of restricting in any way the supply of cereal foodstuffs used in the making of dog-foods. There is no present shortage, but what may happen in the future it is of course impossible to forecast.'

'In practice you will find that only one or two items of your dog's regular diet will become unobtainable and substitutes can be easily found,' said the Canine Defence League. 'For instance, boiled offal or horse-flesh can be substituted for raw beef, and should meat become impossible to obtain, soya bean flour or other protein food can be used.'

Dogs were already getting grumpy at the prospect. It was 'all a bit Mother Hubbard,' to purloin one journalist's clever phrase. Housewives faced queuing for half a morning to get fish heads – or boiling lungs and windpipes (eight hours' minimum) in noisome vats on the soon-to-be declared 'Kitchen Front'. Love of pets in wartime would be true love. It would only get worse. *The Cat* revived a recipe from 1917 for 'a good solid pudding' made from table scraps mashed up with Marmite liquid and baked for an hour in a pie dish into a nutritious cake. It also warned that 'most of the canned foods apart from Kit-e-Kat [made by Messrs Chappie of Slough] are likely to be withdrawn'.

'What Is He Going To Eat Now?' asked Doris Knight in the *Daily Mail* in early October to reassure readers about the prospect of a shortage of pet food. 'Experts are of the opinion that many dogs suffer from overfeeding and that a period of sensible dieting will give most beneficial results,' she wrote. 'Cats will usually eat similar meals to dogs provided there is a moistening of gravy provided by stewing cods' heads or kipper trimmings in water.'

'There's no need to start worrying about how to feed your dog or cat when food rationing comes into force,' she continued. 'The authorities have been giving plenty of attention to the matter for months past.'

Actually they had had other things on their minds. As soon as war was declared, the Ministry of Food had come into being. It was headed by William Morrison, the pre-war Minister of Agriculture, once tipped as a future premier. He would announce on 1 November that rationing (for humans) was to be introduced in the near future – to general grumbling. What about rationing for animals? That was not yet on the agenda but it soon would be.

Feeding pets and farm livestock was not the prime concern of NARPAC. After the catastrophe of the great pet slaughter, there seemed little for it to do. The Committee's affairs were about to be re-energized by Edward Bridges Webb's idea of a mass registration of the nation's pets by 'animal wardens on every street'. To begin with they were to be mustered by the 'Distributing Organisation of the PDSA's Jumble Dept'.

For patriotic animal lovers keen to do their bit, this did not seem very exciting. But create an army of 'Animal Guards' to do the registering and distribute the collar tabs – two million of them – made by ICI out of celluloid, and you had something much more warlike-sounding. Special elastic collars were to be provided for cats.

It was all the idea of a certain Captain T. C. Colthurst, who would later tell an audience how, in September 1939, he was moved to take action after seeing the bodies of dogs that had been thrown into the Regent's Canal in north

London. He was 'Animal Guard No. 1,' he announced proudly. By October there would be thirty of them.

'Animal Guards are Needed Now!' announced *Animal and Zoo News*. 'Who will volunteer? Any adult woman or man over military age is eligible. It is to the Animal Guards throughout the country that we shall turn if our herds of cattle are contaminated by gas. And it is the Animal Guards who will care for our domestic pets if they fall victim to war.'

It was publicist's dream. Pathé Gazette made a newsreel showing guards in swaddling anti-gas suits lovingly rescuing a fashionably dressed suburban woman's Great Dane, which had been 'wounded by shrapnel'. A cosy chat was given on the BBC by Mr Christopher Stone, the honey-toned, Old Etonian Radio Luxembourg disc jockey. 'Do not have your pet destroyed,' he pleaded. 'At the beginning of the war a certain number of people did this. They have regretted it ever since. So far, thank God, the raiders haven't come but you do see, don't you, how important it is that this great plan for the benefit of the animals should be able to work smoothly?'

He mentioned an 'eel, the pet of an elderly lady' and a number of 'mongeese brought to our ports by sailors' as having been candidates for registration – which further 'brought to light all kinds of curious pets, fowls, ducks, goats, lambs, monkeys, small bears, and even a lion cub,' as he would write later.

A news-sheet appeared. Issue No. 2 of the *NARPAC Bulletin* announced: 'A great army of National Animal Guards is organising throughout the country. These voluntary Guards, of which it is hoped there will be one in every residential street, are intended to act as contact officers between first-aid posts and animal owners in each locality. The National Animal Guards will invite owners to register all animals and will provide identity discs. Animal Guards are recognised as doing work of national importance and will wear distinctive armlets.'[9]

All this was months before the Local Defence Volunteers (thereafter the Home Guard) would come into being. Animal Guards? What on earth were they? In fact they were unpaid volunteers with no official status, no preference for petrol, telephones or reservation from military call-up. They did not even have the legal status to destroy a wounded animal without a vet's signature. Nevertheless they would have a rank structure and a six-man (and three women) Grand Council – a kind of pet lover's Supreme Soviet.

According to a rose-tinted, post-war account: 'Offers of help poured in and in the course of a few days over 40,000 people had offered to become National Animal Guards. The process of enrolling them, of arranging them in local groups under Chief Guards, and of fitting such groups into coherent regional organizations, gathered speed. The Guards all wore a white armlet with the now familiar NARPAC symbol, a blue cross in a red circle.'

The triumphalism of late-1939 was extreme. 'For the duration of the war all the animal welfare societies are supporting the National A.R.P. Animals committee,' it was announced in the *Bulletin*. 'It will have under its care at least six million dogs and cats as well as huge numbers of other animals.'

9 The equipment of a National Animal Guard comprises:
One armlet
One house poster
One registration book
100 identity discs and split rings
A number of NARPAC handbooks
Approved adjustable elastic cat collars
One official collecting box, numbered and sealed

There was more. 'The PDSA has placed its Jumble Department at the Committee's disposal. This jumble collection is the *only one* officially authorised on behalf of animals,' it was announced. This was a coup. The RSPCA consulted their lawyers – jumble was charity lifeblood.

The Duchess of Hamilton was equally furious. She had recently made a genteel appeal in the press: 'Who will help by offering a free country home to a dog or a cat belonging to evacuated London people and to those called up for service?' What was wrong with that?

NARPAC had become very cross: this was breaking the common front. They had sent someone round to Animal Defence House to give her a ticking-off. But the Duchess could give as good as she got and told Committee chairman H. E. Dale in December, 'This country has gone to war in the cause of freedom. We strongly resent, as will all earnest friends of animals, any dictatorial attempt to suppress and curtail animal welfare activities.' She knew who to blame for the September 'massacre' as she called it – the Home Office and their horrid 'grey pamphlet' ('Air Raid Precautions for Animals') with its fatal advice.

Furthermore the organization of 'a great army of animal guards' was doomed to be 'as embarrassing a failure as the ARP organisation generally,' Her Grace insisted. That was the Ferne blackout rumpus. Mr Dale replied feebly: 'I assure you we have no dictatorial ambitions.'

So the patrician ark in Wiltshire stayed beyond the socialistic clutches of NARPAC. Something else was stirring at Ferne. 'The officers and men at a wireless station somewhere in England, have owing to the efforts of one of their police staff, come forward with an extraordinary offer to help and several evacuees have been placed with them,' reported *The Dog World* in November. It was all very unofficial. Could refugee pets serve the war effort themselves? It turned out they could.

There had been no rush to recruit pets directly to the aid of the nation's armed forces. Dogs had served the British Army well on the Western Front in 1917–18 as messengers, but in this new, mobile, more technical war what might their place be? Mr H. S. Lloyd, the Crufts champion breeder, was working on a Home Office contract to conduct experiments with police dogs. In November 1939, one of his dogs was tried out as a guard in a test staged at RAF Northolt, the airfield west of London.

A bigger trial followed – with a view to finding suitable breeds and training methods for canines to protect 'detached posts such as RDF [radar] stations where the guard personnel stayed consistent – and to whom the dog would respond and stay loyal'. 'I could get a lot of material from the "Lost dogs homes" and train these,' Mr Lloyd told the War Office.

It was not a success. One dog was 'so fierce it was a danger to the general public, whether innocent or guilty.' Another was so friendly 'it could not be relied upon even to bark at the approach of an unauthorised person'. One dog proved friendly with the men at the guard post but although he 'occasionally barked in the night, it was by no means certain he would so at the approach of an intruder'. This was not going to get the war won.

Chapter 7
Hunting Must Continue!

All the concerns thus far had been about animals in cities. Of course there were plenty more in the countryside – where the nation's agricultural economy was being urgently mobilized for war. Food rationing was coming. What would that mean for animals? Petrol was rationed from the very start. A NARPAC badge was no guarantee of getting any. People must learn all over again how to get around by horse.

Very soon the return of horses to both town and country was generally noted – pulling tradesmen's vans while evacuated townies in the country had resorted to the pony and trap. It would be noted: 'Governesses' carts were getting £40. They could not be given away before the war.'

Urban horses, like pets, were not going to be officially evacuated. Nor could they be taken into shelters. Detailed instructions on how to control horses during air raids were meanwhile put out by NARPAC. Garages and stabling were designated as street shelters plus emergency standing in parks and playing fields in London was scouted out by Captain Hope, the assistant editor of *Riding* magazine.

Colonel Stordy could report on 11 September that the many railway horses in the capital were in good protective order and that the means of removal of maimed or dead animals was in place. 'The King has permitted the use of the Royal Mews at Buckingham Palace as an emergency horse standing, large enough for about 20 animals, and a first aid post,' he said. 'A horse ambulance also is stationed in the mews.'

There was another aspect of equine activity far from the big city that enjoyed royal patronage: hunting. Foxhounds and hunters were not pets but they were civilian animals, inspiring deep sentiment among those who tended them and bitter divisions within the animal 'welfare' world. And so horse and hound (and fox) have their places in this story.

Goaded by a small but vocal anti-hunting lobby, the Masters of Fox Hounds Association had made soundings in early 1939.What would war mean for hunting? Would there be rationing, or mass mobilization of horsepower for the Army? A public opinion-winning move would be a voluntary reduction in dog packs (expressed as 'couples' – pairs of dogs) and oat-consuming hunters.

The MFHA circulated members on 29 August with instructions 'in the event of war', while admitting, 'cub-hunting [the season had begun on the 4th] may only be carried on under very great difficulties. But it would be prejudicial to the country to stop it altogether.'

They recommended that 'Cubbing should take place where conditions allow, in order to kill as many foxes as possible, but it should not be looked on as a form of sport as long as the war lasts.' Meanwhile masters should, 'consider reductions in their establishments generally.'

*

When war came, *Horse & Hound* magazine had no doubts. 'Hunting must continue!' it proclaimed on 22 September.

With men being called up, the sporting paper insisted: 'Women huntsmen will carry on so that, after the forces of evil have been run to ground, we can continue once again with the traditional sport of our fathers.' It was suggested that evacuee children should follow hunts and visit kennels to see for themselves the importance of hunting in country life. 'War or no war,' reported *The Tatler*, 'it is on record that the Quorn have been killing their fox a day.' A page of photographs depicted a special day's hunting organized for the officers of the Life Guards.

The Field lamented, 'War is upon us at the very commencement of cubbing.' The magazine's correspondent out with the Croome for its first austerity hunt of the season in November 1939, noted: 'No brave scarlet and gleaming toppers, instead rat-catcher interspersed with khaki, unclipped horses and a sadly depleted field.' The writer however knew what mattered: 'But let us forget the troubles of this mad world. Wars and rumours of wars retreat into insignificance when hounds are running.'

Hunting was also important in British military life. The place of the horse-borne pursuit of foxes in modern war was harder to define than in 1914 when it was both the cradle of valour and a source of cavalry mounts. Soon after the outbreak of this new war, the 1st Cavalry Division had been sent off to Palestine to do nobody knew quite what. The British Expeditionary Force had been shipped across the channel propelled entirely by internal combustion engines plus a number of mules. By late 1939 its men were coming home on leave. Would they bring back adopted foreign pets?

The Ministry of Agriculture reminded everyone how smuggled dogs had caused serious rabies outbreaks at the end of the First World War and there should especially be no regimental pets belonging to 'colonial or dominion troops, which had been the source of so much trouble in 1914'.

It posted a stern notice: 'Soldiers and airmen are reminded of the dangers of allowing stray dogs and cats to attach themselves to them. They must be handed over to military police for disposal.' And early in 1940 there was an Air Ministry order amending King's Regulations, 'to prevent the movement or importing into this country by air of dogs and cats'. There must be no flying pets.

In the same British Expeditionary Force however, now in France, there were reportedly officers keen to keep up the old traditions by importing packs of hunting dogs from Britain, as Wellington's officers had done. The French Minister of the Interior refused to make any of the countryside available. He told them, a little coolly, that the French treated the war seriously.

The Royal Artillery Foxhounds, with the exception of seven couples, were destroyed soon after the outbreak of war, but one hunting historian states: 'Major Montacute Selby-Lowndes took a pack of beagles to France with the British Expeditionary Force,' while the whipper-in of the Royal Artillery Hunt, Captain Frederick Burnaby Edmeades, 'managed to smuggle a couple of harriers with them to France and enjoyed several weeks hunting until apprehended by the Gendarmerie and hauled before the Army Commander.' All very dashing but it was not modern warfare.

When an MP asked in Parliament on 2 October whether the Minister of Agriculture 'should take advantage of the present situation and bring this savage and destructive sport to a complete close?' it was a marker of the way things were going politically. The chairman of the British Field Sports Society wrote the next day to the agriculture minister, Major Sir R. H. Dorman Smith, to say: 'Where would the eighteen or so mounted yeomanry regiments found today get their horses were it not for hunting? Mounted regiments may yet come into their own in this war.'

Lord Burghley, Master of the East Sussex Hounds (then a junior minister at the Ministry of Supply), wrote to the Minister on the 4th: 'What I hear from fellow masters is that they are making large reductions on their packs to keep the show going until the end of the war so that one of the greatest and happiest facets of our country life may not be lost and centuries of careful breeding destroyed in a moment.'

The story of field sports and war was to be a tortuous one. Lord Burghley would have a special place in it.

Of course the countryside was much more than a sporting playground. It was where food came from to feed the nation – or at least some of it. Half of the bulk feedstuff for Britain's cattle was imported. It had long been cheaper than home grown. The sea lanes were not yet contested, but they surely would be, as they had been by German submarines in 1917–18, when greedy dogs were blamed for eating all the food. A prominent Tory MP had declared if he had his way, he would have 'every Pekingese dog in the country killed and made into meat pies'.

This time round planners knew that, should war come, importing killed meat in refrigerated ships was a more efficient means of getting protein into the country than importing grain to fatten British herds. Sheep meanwhile ate British upland grass.

To survive the siege there would have to be fewer cattle in the nation's fields, but eating more home-grown feedstuff coming from more land, which had yet to be put under the plough. There was nothing to spare – not even chickenfeed.

Every source of protein was vulnerable. Eggs came from Poland, even China. 'It is too early yet to frame a definite policy in respect of pigs, poultry or eggs, in view of the large amount of cereals [maize and barley], required for these forms of production, a large proportion of which has in the past been imported,' the Agriculture Minister Reginald Dorman-Smith announced on 19 October 1939.

But there had to be some sort of action. There were sudden local shortages. Foodless pigs and chickens were being slaughtered wholesale. By the winter of 1939–40 there was generalized panic that the larder was emptying. Anything with a mouth or beak that could not itself be eaten, even those that could be, seemed to be an enemy within. Animal-loving Louise Lind-af-Hageby despaired at the 'stupidity' of it all when she wrote a little later:

> Those clamouring for the killing of animals knew already that the number of cattle had been severely reduced, that hens had been killed on an enormous scale, there had been a clamour for the trapping of ten million moles, that the pigeons of cathedrals and Trafalgar Square had been threatened with extinction. In some people's minds, the idea of winning the war had become associated with exterminating every non-human creature.

Crows, jackdaws, sparrows, starlings, pigeons and rooks were being trapped or shot in their thousands. For the lowlier rural orders, according to one authority, 'the war seemed a splendid excuse for the legalisation of poaching'.

When partridge and pheasant shoots became unviable (beaters had been called up), *The Field* magazine suggested hunting grey squirrels, which had been declared a pest under 1939 emergency legislation. But not necessarily to eat:

> Shooting men can get a deal of fun in dealing with him. And in helping to rid the country of an animal in whose favour it is difficult to say anything, they will be doing work of national importance.

The war might be sundering urban humans from their pets, but it was filling the cities with different sorts of animals. There was the return of the town horse. Now it was animals you could eat. Soon after the outbreak of war, the Minister of Agriculture had appealed for an army of backyard poultry keepers to be raised. As *Eggs*, the magazine of the Scientific Poultry Breeders Association, declared on 6 September: 'We appeal to every reader of *Eggs* to be of good heart and confident in the future,' while admitting, 'the interference with grain supplies is likely to cause us much trouble.'

The writer George Orwell was a keen reader of *Eggs*, as a smallholder at his tiny country cottage in Wallington, Hertfordshire, some twenty miles north of London. His winter 1939 diary recorded plans for springtime poultry breeding and 'maybe going in for rabbits & bees'.

'Rabbits are not to be rationed,' he wrote. 'The butcher says that people will not as a rule buy tame rabbits for eating but their ideas change when meat gets short.'

Rabbits were going to have an interesting war. Wild, Mr McGregor-taunting, crop-munching country rabbits were the nation's enemy. Farmers could blast, gas and trap away as they pleased. Tame, fat city rabbits being fed on cabbage leaves in backyard hutches, however, might be the nation's salvation. How else could one obtain 2½lbs of meat for feeding costs of no more than 6*d*?

Three does would produce an average of sixty youngsters in a year, readers of *Fur and Feather* were informed, but it would be better to have four. The mating must be spaced out, lest all the litters arrived at the same time. Some householders found that tenancy agreements forbade them to keep rabbits, but in the emergency of mid-summer 1940 all restrictions would be lifted by the Government. Soon there would be rabbits everywhere.

On 1 November a brave new magazine appeared: *Goats*, edited by Mr W. O'Connell Holmes of 'Capricorn', Polstead, Suffolk. 'Each and every goat dairy in the country is now a vital stronghold in the defence of the Home Front,' he declared stirringly. Under the heading 'ARP for Goats', he advised: 'To prevent panic in the goat house from the sound of bombs, give your goats, especially the highly-strung milkers, a sedative medicine.' His journal would prosper.

The Pigeon Fancy was having an altogether tougher time as the Government wanted message-carrying birds for its own purposes. Strict wartime regulations aimed to both control birds in private ownership (lest they bear treacherous messages out of the country) and to clear the skies for official pigeons.

It became an offence for anyone to keep pigeons without a permit, to free pigeons, or to kill or wound any racers or homers. Anyone finding a pigeon, alive or dead, with an identification mark or a message attached, was to hand it over at once to a policeman, making no attempt to read or decode the message. Farmers and sportsmen were urged to take special care not to shoot long-distance pigeons – you never knew what they might be carrying.

Bee keeping was also very patriotic. Bees might have a role beyond the production of honey, too. An article in *PDSA News* would suggest:

> Bees as messengers – bearing tiny rolls of paper they would be better than pigeons and their drone more

> reliable. Coloured by powder according to a pre-arranged signal, they could carry urgent information safely, and speedily.

The War Office, sadly, was not interested in the proposal.

Chapter 8
Wolves Not Welcome

Britain had gone to war for the sake of Poland. That country was now long beyond help. The animals of Warsaw Zoo had been bombed on 25 September 1939, two days before the city's surrender, the first of many European menageries to be smashed to pieces. The seals escaped into the River Vistula, ostriches and anteaters roamed the Old Town. Cristina Zabinski, the director's wife, kept a diary. 'Submerged in their wallows, the hippos, otters and beavers survived,' she wrote. 'Somehow the bears, bison, Przewalksi horses, camels, zebras and reptiles survived.'

Pretty soon a visitor arrived. Lutz Heck, the director of Berlin Zoo, smiling and persuasive just as he had been as a colleague and collaborator in the pre-war days, engaged in his strange quest to recreate the beasts of Neolithic Europe by selective breeding. He had joined the Nazi Party in 1938 and found high-level backing for his plan to populate the primeval Białowieża Forest in eastern Poland with eugenically recreated ancient aurochs cattle and tarpan horses.

Warsaw Zoo's survivors, including the baby orphan elephant 'Tusinka' (whose mother had been killed in the air raid), were smartly carried off in special trains to Germany.

Heck's career had been extraordinary. The most significant zoologist in the world in the decade before the war, to mark the 1936 Berlin Olympics he opened a 'German Zoo' – an exhibit honouring the country's wildlife, complete with 'Wolf Rock' at its centre (an enduring feature of Teutonic animal parks). But now Herr Heck had concerns closer to home. His autobiography recorded the summer of 1939 as war approached and Nazi officials, like those in London, became concerned with what to do should any of Berlin Zoo's 4,000 animals escape in an air raid. Lions and tigers would seek shelter rather than attack humans, so Herr Heck pleaded, and snakes (except the African mamba) would be numb and sluggish without artificial heat.

A 'frightened elephant' however was very unpredictable. The most dangerous animal in fact was the German stag (in the rutting season). Nazis and pets presented a paradox. Where, in a political system erected on racist biological abstraction, did pets figure?

They were seemingly at the heart of things. Germans loved their pets as much as anyone else; it was other people they were not so sure about. The Third Reich's strict animal protection laws, enacted in November 1933, had taken welfare of all animals – domestic, wild and commercial – into a new dimension. But pets were highly politicized. And they were culturally ambiguous. Pets were bourgeois. Pets were tame. Some dogs might be honorary wolves but cats seemed wilfully disinclined to obey orders.

The welfare laws themselves were also politicized, underpinning the anti-Semitic goals of the regime in expressing 'healthy German popular sentiment' against kosher slaughter and animal experiments performed by 'Jewish' science. These laws also banned fox hunting by what was left of Germany's toffs.

The Reichstierschutzbund (the Reich SPCA, for want of a better description) absorbed local animal societies run on cosy British lines with little apparent fuss. Its headquarters were in Frankfurt-am-Main, with the city's Nazi mayor, Friedrich Krebs, as its 'leader'. From 1938 membership was restricted to those of 'German or related blood'.

Detailed 'welfare' laws continued to be decreed almost until the very end of the Third Reich, concerning such matters as the 'humane' killing of lobsters, or how to transport a dog on a railway train. The irony of such Nazi concern for animals, when human life was held in murderous contempt, has often been remarked upon.

Dogs however had a special place in the national socialist state. Boxers, German Shepherds and certain other breeds were militarily useful. When the German Army went to war in 1939 it already had thousands of 'skilfully trained sheepdogs', so it was reported in London. In December it was decreed that dogs measuring more than 50 cm to the shoulder should be registered for potential war service with allotted rations of offal and oatmeal. But they would have to pass rigorous tests to get it.[10]

10 An anonymous Internet memoir tells the story of a twelve-year-old in Linz, who took his German Shepherd, 'Donar' – described as 'a pure-bred dog with a first-class pedigree' – for the obligatory assessment. His friend Willi's Saint Bernard, 'Barry', was there too. They had contrived between them to make sure their pets failed the test so as not to lose them to the military.

'The day of the dog evaluation came, a beautiful hot summer day. Willi and I went with Barry and Donar to the meadow behind the Bulgariplatz where the dogs were to be evaluated. A man in civilian clothes made a speech. He said we must all be proud that our dogs might fight for our leaders, our people and country. Two soldiers fired their guns simultaneously. At once Barry scurried loose. I kicked Donar hard without them noticing and he yelped and ran.

'A couple of days later my parents received a registered letter. It stated that our Donar was found to be unfit for service. Any German

For less warlike animals, 'special ration cards will henceforth be issued for dogs of exceptional pedigree, sporting dogs and dogs of the blind,' so British newspaper readers were told, 'which will entitle them to an allowance of oat or barley meal. The rest are to be destroyed to conserve food supplies.' This in fact was a propaganda lie. 'Cats are exempt from destruction as they keep down rats and mice.'

In *Animal and Zoo Magazine* the writer Carl Olsson recalled the stirring events of the blockade starvation winter of 1917–18, 'when ordinary Germans who had always loved their zoo and their academic leaders who had made them into the finest institutions of their kind in the world, protested at attempts to cull the animals – and thousands were saved to see the Armistice.' Not this time. The ferocious beasts of Berlin were doomed, although Hermann Göring had rescued some lion cubs for his hunting estate. And this was not just propaganda about Nazi frightfulness. Lutz Heck's post-war autobiography described air raid precautions and gas drills in September 1939 as he toured the garden with his Wire Haired Fox Terrier 'Lutting'.

'Orders were given for all beasts of prey to be shot,' he wrote. Very soon all the lions, tigers, leopards and bears were killed.

It was further reported in London that the 'Reich Minister of Agriculture had ordered all elephants, yaks, camels and oxen that might serve as draught animals to be registered' and sent to a special school near Munich, where 'circus trainers would break them in for war work.'

Shepherd that is unworthy is not breed-worthy and therefore is not allowed to be used at stud for any German bitch. We were commanded to return the pedigree certificate, to be destroyed by those responsible for the purity of the breed.'

Hamburg Zoo and Hagenbeck's travelling menagerie were doing the same. Elephants were pulling ploughs in Hanover and hauling lumber in the Black Forest. Hamburg Zoo's lions, tigers and all carnivores had been shot out of hand, so British animal lovers were informed. 'The fish eaters, seals, walruses and sea lions, pride of Berlin Zoo, have all been destroyed, their blubber utilised and their meat put on the menus of Berlin restaurants. All reptiles have been destroyed.'

Dresden Zoo's lions, tigers and panthers had all gone the same way. All the snakes had perished except the boas and pythons, one of whom was fed an entire goat just before the outbreak of war. Munich Zoo had managed to evade the killing order, apart from 'a few of the bigger chimpanzees'.

The Paris Zoo in the Bois de Vincennes had shut for a week, like Regent's Park, when mothers and children left the French capital. The majority of the animals were evacuated south and west. 'Some went on loan to travelling circuses, even the fish that went to private aquaria, including the one owned by the Prince of Monaco at Monte Carlo,' so *Animal and Zoo Magazine* reported. 'Only a few animals were killed, a morose and dangerous orangutan and an obstreperous bull elephant that would not enter his travelling box and had to be shot. Some wolves were shot as being generally unwelcome as guests.'

London Zoo took steps to advertise the fact that it had reopened. Indeed, so it was reported in early October, 'only a few redundant and elderly animals have been destroyed'. Come back to the zoo, 'the animals will enjoy your company' was the message – 'and bring stale bread and buns, Spanish chestnuts and nuts of every kind, the outer leaves of cabbages and lettuces to feed your Zoo favourites.'

Heart-warming stories were back in the newspapers – for example, the one about 'George' the chimp, who was doing patriotic tasks such as knitting socks for soldiers. Jerboas, chinchillas and echidnas were all said to be enjoying the blackout. The outdoor leopards were, 'interested in watching the rise and fall of the balloons which are a source of never-ceasing wonder to them'.

On 13 November, 'Sammy', a runaway black genet on the loose for thirteen months, was found dead in a trap in Romford. The following day, 'Koko', the Zoo's most famous chimp, died at Regent's Park.

Twin African crested porcupines were the Zoo's first war babies, of which only one had survived. The tiny urchin was in the Small Mammal House, being kept warm by an electric lamp. What would the momentous events engulfing Europe hold for him?

Part Two

THE MINISTRY OF PETS

I am sorry, my little cat, I am sorry –
If I had it, you should have it;
But there is a war on.
The butcher has no lights,
The fishmonger has no cod's heads –
There is nothing for you.

'To Her Cat in Wartime', Dorothy L. Sayers, 1940

Chapter 9
Pets Get the Blame

There was little that was phoney about the 'phoney war' for pets. Even if no real land fighting was going on, in the winter of 1939–40 the perils of the blackout, evacuation and an unpredictable diet made the months following Poland's subjection an uncomfortable time for Britain's Home Front animals. It was about to get worse.

On 8 January 1940, bacon, butter and sugar were rationed. Meat would surely be next. It was getting scrappy round the food bowl. The enormous proportion of imported food, especially meat and animal feed that came by ship to the as yet only mildly embattled island, made national survival look perilous. The public were only just waking up to the danger as the 'Dig for Victory' campaign sought to turn every garden into an allotment and every backyard into a miniature livestock farm. The trouble was, these were pets you were meant to eat.

On the first day of the New Year the National Poultry Council had expressed 'grave concern at the large quantities of feeding stuffs which are being consumed by hunters, packs of hounds, dogs, pigeons etc, not engaged in war-work'. The whispering about 'useless mouths' was growing louder.

Corn imports for chickenfeed had already been cut to a trickle. 'Even dogs are given more consideration than poultry,' an outraged chicken farmer declared at the start of the year – 'there is no restriction on the manufacture of dog biscuits made from just the ingredients required for egg production.'

Evacuated London children had meanwhile turned from being lovable urchins into nit-ridden horrors. 'I've seen a dog with better manners,' pronounced a shocked matron in the shires. Displaced townie dogs were being blamed for a wave of sheep worrying as lambing time drew on. Dog lovers generally sniffed a growing anti-canine sentiment. Anyway the New Year was always a bad time to be a British dog: it was when the *7s. 6d.* licence fell due. As Mr Slee of the Dog and Cat Infirmary, Plymouth, told the local paper: 'It was always found after January 1 that strays became plentiful. Bitch puppies in particular seemed to be unwanted, and were carried in baskets to the city outskirts and turned adrift.'

'The Market was also a common dumping place for these castaways,' said the reporter. 'After running about for days, collarless and uncared for, they were picked up by the police and others and taken to the home. Unless they took the fancy of people in search of a pet, the fate of these animals did not appear to be very promising.'

Cats fared no better. 'Although not affected by the licensing, they were even more liable than dogs to find themselves without a home,' said Mr Slee. 'It was invariably the young female cats which met with that treatment. It was just a case of not being wanted.'

'The black-out had resulted in fewer pets straying from home,' thought Miss Amble of the West Country pet shelter. Just after war had been declared, she was kept busy destroying pets 'for people who thought they would be unable to obtain food for them,' it was explained. 'Since then, however, business at the home has been quiet.' (Plymouth would be pounded by the *Luftwaffe* in spring 1941.)

Still no bombs had fallen. As *The Animals' Defender*, journal of the Animal Defence Society, noted at year end: 'This curious war that never begins (except at sea) is at any rate affording time for ample preparations.' They meant NARPAC and its mission to see a blue-cross-badged collar at every pet's throat.

The Committee confessed in a New Year statement: 'The aim of having an Animal Guard for every street is a long way yet from achievement. There has, however, been a good response to the Committee's appeal for the registration of animals. Completed forms are arriving daily in hundreds.' It had set up offices in Manchester and Birmingham – and was planning a grand move to take responsibility, not just for urban pets, but for all animals of 'economic value'.

An extension of its activities nationwide would be a coup. The National Veterinary Medical Association (which was represented on the committee) pointed out that it was still completely unofficial. Would it not be better all round if it got recognition – with the crown device on its badge – just like the proper ARP?

Some people were getting jealous. 'NARPAC Caught Napping?' *The Animals' Defender* asked in March. 'Complaints are coming to us from so many independent sources that we cannot suppose they are all entirely baseless.' There were multiple complaints about no one at the HQ acknowledging offers to be Animal Guards and a lack of collars, according to the Animal Defence Society journal: '[It could] unhappily end in the whole scheme having to be scrapped for want of co-operation from a disgusted public.'

Louise Lind-af-Hageby complained that the celluloid discs were 'highly inflammable and a danger not only to animals but human beings'. She got this reply: 'A cat or dog lying in front of a fire must move long before the disk attached to it catches fire.'

The turmoil of evacuation (and by now a mass return to unbombed cities) and called-up dads still caused heartbreak. On 14 February the *Daily Mail* highlighted the plight of four-year-old Golden Retriever 'Neil', who been evacuated to Aberdeen with Gillian, the young daughter of its owners, Captain and Mrs R. H. Donald. The dog, more used to the modern comforts of Dolphin Square in Pimlico and garrison life in Aldershot, had vanished from its temporary Scottish billet, apparently 'looking for its master who is now serving in the British Expeditionary Force. Gillian is said to be fretting for Neil.'

Fathers in the services and mothers on war work meant a widespread concern on the part of canine worriers for a supposed large number of 'lonely dogs'. Volunteer neighbours should at least offer to take them walkies.

Then there were happier tales. The Our Dumb Friends' League's Paddington Shelter reported 'five white ferrets brought to the shelter because the owner was leaving for the Army and felt they would not be happy with anyone else'. And a number of Angora rabbits were rescued after delayed action bombs had been exploded or made safe.

In March 1940, *PDSA News* told the story of the Revd M. Duke, who had left the living of Saint Mary's, Doncaster, for Emmanuel, a church in Paddington in west London. 'His cat, Whiskers, who had accompanied him to London turned up in Doncaster ten weeks later, where it had left a number of kittens,' said the report. 'The only explanation I can think of,' said the vicar, 'is that cats have some power of distant communication.'

There were plenty more stories about evacuated cats with amazing homing instincts, including a ten-year-old tabby who voyaged from Saffron Walden to Surrey, another who got from Devon to Surbiton, and yet another who travelled from London to Brighton guided by some impossible feline instinct.

'Peter', a three-and-a-half-year-old Siamese, belonging to Mr and Mrs Jenkins of Surbiton, walked home from Devonshire, a distance of 187 miles. It took him three months, 'but he arrived safely on the front-door steps of his home'.

'Rota', a pet lion from a circus 'won in a bet' as a cub by its owner, Mr George Thompson, managing director of Rotaprint, lived in a substantial cage in the garden of 49 Cuckoo Hill Road, Pinner, in suburban north London. Neighbours complained about the roaring. ARP Wardens demanded to be armed. Rota ate 50 lb of horseflesh a week. On 31 May he went to the London Zoo. His adventures were just beginning.

A posh cat had a narrow escape via human intervention. In spring 1940 in as yet unbombed London, two office girls in their lunch hour found 'a frightened tabby' wandering in Curzon Street, Mayfair. They took it to a shop in nearby Lansdowne Row run by a Miss Marjorie Ashton, who had an 'Animal Guard' sign prominently displayed in the window.

Other than being in obvious immediate distress, the tabby was evidently well cared for. It was fortuitously wearing a NARPAC blue-cross-red-circle disc on its collar inscribed with a reference number – which, after some urgent telephoning, quickly divulged an address not too far away in fashionable Eaton Place, Belgravia.

The fortunate tabby seemed all set to be reunited with its upstairs owner but at the tradesman's entrance of No. 47, the downstairs housekeeper refused to acknowledge it. The family cat, she insisted, had already been 'despatched in a hamper' by railway van to its mistress, Lady Juliet Rhys Williams. She was the daughter of the romantic novelist Elinor Glyn and the wife of a prominent Liberal politician, herself an ardent social reformer. Once described as the 'cleverest woman in England', Lady Juliet was not clever enough to hang on to her cat.

Lady Juliet was apparently taking refuge at the family estate at Miskin Manor, Glamorganshire. The distressed tabby must have somehow escaped from the 'hamper' between Belgravia and Paddington station. It was on its way to Wales in a more robust container the next day.

The wider war was not entirely forgotten. In March 1940, the fate of the gallant Finns, having resisted the Soviets for five months, grabbed the attention of British animal lovers.

The RPSCA had sent the intrepid Lt-Col Gartside to the Baltic and opened a fund for the animals of 'tragic and unspeakably heroic Finland as they faced the shells, bombs and bullets of the cruel Soviet masses'. Nothing was worse than mistreating animals. Russian horses captured by the Finns, it was reported, 'showed signs that their normal rations were anything but satisfactory'. There were tales of the Russians training dogs as canine mines to blow up Finnish tanks.

Madame de Gripenberg, wife of the Finnish minister in London, appealed on behalf of 'the faithful little Finnish Spitz dogs, so clever as messenger carriers and in their ability to find their masters in the snow, the small horses, sturdy, swift and intelligent, the reindeer used to draw sleighs of wounded men.

'We hope and believe that even in this period of financial stringency there will be many who will find it in their hearts to help the RSPCA to help the animals in Finland.' By the time these words appeared in print, the exhausted Finns had already sued for peace. There would be plenty more of Europe's animals to succour in this war.[11]

Concern for animals began at home. The Government was shaping up for a long siege. That spring 5,000 civil servants of the Ministry of Food were decanted to the faded Edwardian seaside resort of Colwyn Bay in north Wales, to fight the good fight with memo and rubber stamp from requisitioned hotels and boarding houses. The Meadowcroft Hotel on Llannerch Road was to be the headquarters of the Animal Feeding Stuffs Division. It would function, *de facto*, as the Ministry of Pets.

Meat was declared rationed on 11 March 1940. It was done by price rather than weight, to the value of 1*s*. 10*d*. per person per week. Cheap cuts became premium cuts. For carnivores, it was to be a tough time. Sausages, of dwindling meat content, were not rationed and nor was offal (liver, kidneys, tripe, oxtail, kidney, heart etc.) but that did not mean you could get it.

Tinned pet food such as Chappie and Kit-e-Kat were not restricted (yet). *The Cat* urged the vegetarians among their readers to donate their physical meat ration to cats' shelters.

Advice crowded in from all sides on the feeding of pets, from newspapers, from welfare groups, from

11 A strange story surfaced in 2011 from wartime German files about a Dalmatian called 'Jackie' owned by Tor Borg, a Finnish businessman. In late 1940 his German-born (but anti-Nazi) wife, dubbed the dog 'Hitler' because of the way it raised a paw in the 'German salute' whenever the Führer's name was mentioned. News reached Nazi diplomats in Helsinki, who were not amused and sought ways to bring Borg to trial for insult.

manufacturers. NARPAC had issued a booklet, *Wartime Aids for Animal Owners*, which stressed a balanced diet for cats and dogs, containing its proper quota of 'energy-giving', 'body building' and 'protective' foods. A large Pekingese weighing 10 lb needed 5–6 oz of carbohydrates, 1 oz of protein and a ¼ oz of fat at each meal, while for a 25 lb Scots Terrier and a 60 lb Airedale the amounts had to be proportionate.

'This country has not reached a stage when the wholesale destruction of household pets is necessary,' counselled the RSPCA. Its handy leaflet *Feeding Dogs and Cats in Wartime* advised, 'Potatoes are plentiful and if you put in extra tubers when digging for victory you will not have it on your conscience that shipping space is being taken for food for your animals.' Starchy potatoes could be made palatable with gravy concocted from bones. Perhaps.

Other suggested canine diets consisted of stale bread and oatmeal, 'made into thick porridge' and mixed with meat scraps from the table, or stale bread mixed with 'chopped cabbage, cauliflower, brussels sprouts, turnip, carrot or other green leaves' moistened with 'soup or gravy made from bones or scraps'. Gravy from to-be-human-consumed stewing meat could be offered to pets, 'a sacrifice that every owner would be prepared to make'.

Proprietary foods, 'Red Heart, Kit-e-Kat, Ken-L Ration, Chappie etc.' were available in glass bottles, said the Society, although being a charity, it could not make recommendations.

Cats, the Society advised, 'will usually eat the food recommended in the above diets, provided there is included some meat gravy, sardine oil, or oil liquids made from fish trimmings. The Ministry of Food will not permit the use of Cod Liver Oil as food for household pets.'[12]

The pacifist Louise Lind-af-Hageby meanwhile could find comfort in how the war had inspired 'the majestic rise of the vegetarian life, once supposed to be chosen by sentimentalist weaklings and the gastronomically foolish. The potato and the carrot have assumed enormous importance.'

Too much so for some ... The Cabinet Food Policy Committee decided at the beginning of May that impending shortage was going to mean reductions of foodstuff all round for 'non-essential' livestock. Horse racing should be reduced and greyhounds limited to one meeting per week per track as a statement of intent, if nothing else. Hunts would be rationed to one sixth of the pre-war level of feed for hounds (including beagles) and one tenth for hunters.

Pig and poultry keepers were already getting one sixth of pre-war levels. Game and silver fox farms were strictly rationed. Zoos were severely curtailed – one third of pre-war consumption for London Zoo, one fifth and later one tenth for private zoos of pre-war levels. The hunt for horseflesh for the carnivores was constant. Getting fish was 'very difficult'. Two 'big hungry sea lions' were shipped to America. Carrots for fruit eaters were grown at Whipsnade. An adoption scheme was launched, initially for ZSL Fellows, rapidly expanded to anyone who wanted to send money for food – typically hard-to-get but unrationed fruit.

12 The prescription of free 'welfare foods', cod liver oil and orange juice, for children and expectant mothers was introduced in December 1941. It was a serious offence to give it to cats but some mothers (but not all cats) found it hard to resist.

'Adolf' the aardvark (thus named as a baby in 1936) was one of the first to be adopted. He was renamed 'Charlie' but people seemed to prefer his earlier incarnation. Dorothy L. Sayers, the animal-loving crime writer, adopted a porcupine.[13]

Wheat for dog biscuits was reduced by two thirds of pre-war amounts. The 'Milled Wheaten Substance (Restriction) Order' made on 2 May, biscuit-loving dogdom's bane, forbade the use of such products for anything but human food without special licence.

The National Milk Scheme would be launched on 1 July, to provide expectant and nursing mothers and children under five with daily milk at a fixed price of 2*d*. a pint, or, if necessary, free. Otherwise for adults it was two pints a week. There was nothing spare for pets.

Hungry humans knew who it was to blame. Anti-pet whispering became a clamour. Grumpy farmers were already advocating the destruction of domestic pets – while their own working dogs were exempt from licence. Tail-Waggers advised that there were 'dreadful people' abroad leaving lumps of poisoned bread in the streets and advised dog lovers to carry a lump of washing soda in their pockets ready to push down a poisoned dog's throat.

How to reduce dog numbers was becoming a direct concern of the Government. Raising the cost of the dog licence was considered in Whitehall in May. But it was noted by HM Customs and Excise for the Minister that this 'would fall on the lower classes' who fed scraps to their

13 Which at 1*s*. a week was one of the cheapest creatures to feed. A lion cost 15*s*., an elephant £1. Most expensive were the sea lions, some of which were sent to America. Visitors were further encouraged to bring kitchen scraps, acorns and stale bread. Such measures in effect kept the London Zoo going through the war.

The National Archives

State Library of Victoria

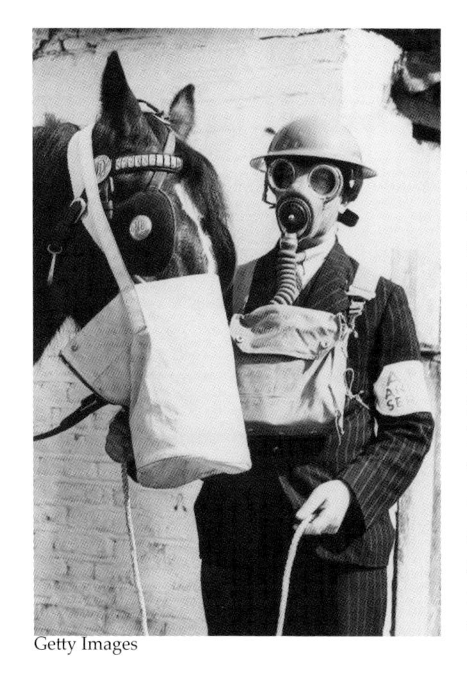

Getty Images

Fear of a devastating mass attack on Britain's cities at the outbreak of war drove Government policy on Air Raid Protection. Would pets go barking mad at the first wail of a siren? All sorts of 'gas-proof' containers were touted (*above*) while in early 1939 Mr. Harold Bywater (*left*, with friend), chief municipal vet of West Ham in East London, devised a pilot protection scheme for the dockland borough's many thousands of domestic pets and working animals.

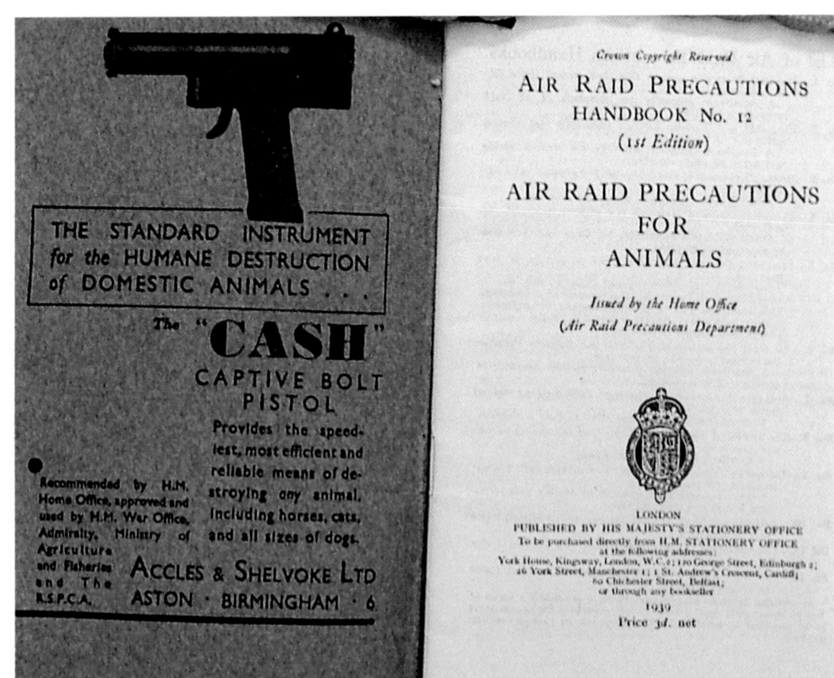

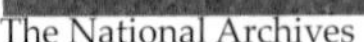
The National Archives

Christy Campbell

YOUR DOG AND CAT IN WARTIME

CONTENTS

IN the strenuous and trying days which face us all, countless owners of dogs and cats will get from their pets a measure of faithfulness and companionship which may well be a great comfort to them.

In this booklet we show how you can best care for the welfare of your dog or cat in wartime; and if we can help you on some particular problem of your own, and which is not covered in this book, we shall be very pleased to do so.

Bob Martin

FOUNDER, BOB MARTIN, LTD.

Wandsworth Heritage Service

Dog-Owners!

You are not allowed to take your dog into a Public Air Raid Shelter–

but

BOTH YOU AND YOUR CANINE FRIEND ARE WELCOME HERE WHEN A WARNING IS GIVEN

MUZZLES ARE DESIRABLE

NO RESPONSIBILITY CAN BE ACCEPTED

For Advice on any question affecting Dogs, ask

THE NATIONAL CANINE DEFENCE LEAGUE

VICTORIA STATION HOUSE
LONDON, S.W.1
(CHARLES R. JOHNS, Secretary)

H. L. Castle, Ltd., Water Lane, Watford.

Sussex University Mass Observation Archive

Christy Campbell

Imperial War Museum

The announcement of the existence of NARPAC (the National Air Raid Precautions Animals Committee) on the eve of war did little to reassure pet owners who had been officially advised in Home Office ARP Handbook No. 12 (*opposite, top*) that to have their animals destroyed was the 'kindest thing to do.'

The rendering company, Harrison, Barber & Co. of Sugar House Lane, West Ham (*opposite, left*, its yard today part of a huge post-Olympic regeneration site), was overwhelmed by the massacre of more than 750,000 pets. The Bob Martin pet health company (*opposite, right*) and the National Canine Defence League (*above left*) advised very differently – while NARPAC's registration scheme offered some semi-official comfort that pets could survive. Here, in an early 1941 image, wartime housewife Mrs. Olive Day (*above right*) shows off her delightful black cat, 'Little One,' with NARPAC collar.

German pets found protection in the Reich Animal Defence League (*above*) while cats found a champion in Prof. Friedrich Schwangart's 1937 book *On the Rights of Cats* against charges of being 'poachers.' They were 'hygienic helpers' in the war on mice, said the Professor.

Pedigrees and mongrels alike were caught in the great killing panic of 1939 but, to begin with, posh pets had a better chance of survival. The Royal corgis were evacuated to Windsor where they had a cottage in the Great Park.

Fashionable stores promoted luminous blackout jerkins for dogs (*right*), while an opportunist breeder exhorted owners 'to carry a white Pekingese' to avoid accidents on the darkened streets.

Getty Images

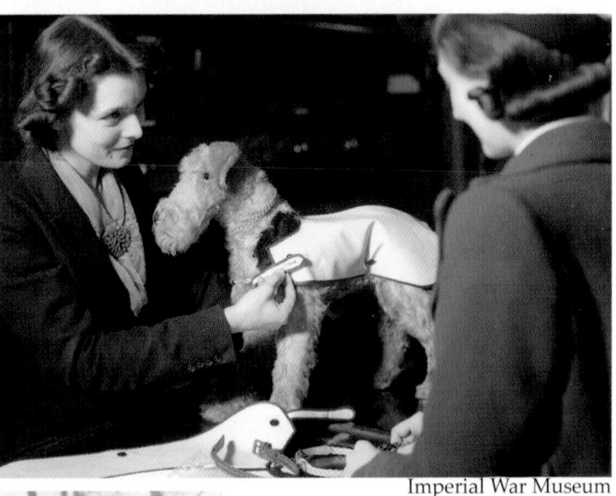

Imperial War Museum

Imperial War Museum

The National Canine Defence League sponsored an air raid shelter in Kensington Gardens where brave but foolish exercisers had been routinely observed not taking cover on air raid warnings.

National Portrait Gallery

Christy Campbell

The remarkable Duchess of Hamilton declared her London and country homes to be sanctuaries while her campaigning HQ, Animal Defence House in St. James's Place, was full of refugee pets. Her patrician rescue operation was however aimed at London's poor and the better-off were politely told to make their own arrangements. MI5 meanwhile described the Duchess as a 'well known crank' (*below*).

The Duke's mother is a well known crank Animals' etc!

The National Archives

After months of 'phoney war,' in May 1940 the Dunkirk debacle saw hundreds of dogs (and a few cats) crossing the Channel with the bedraggled British Expeditionary Force.

When the Blitz began in September, to general surprise, pets stayed calm under fire. Below, London firemen relax with their splendid poodle mascot.

Getty Images

Imperial War Museum

The work of pet rescue largely fell on animal welfare charities - like this extemporized PDSA post in Plymouth in early 1941 (*top left*) and this National Canine Defence League clinic (*below left*), 'more open than ever' after itself being hit by a bomb.

An army of mostly female amateur rescuers did what they could in the rubble (*below*) – this young woman in West London following NARPAC advice on the use of an extemporized grasper.

Western Morning Post via Derek Tait

State Library of Victoria

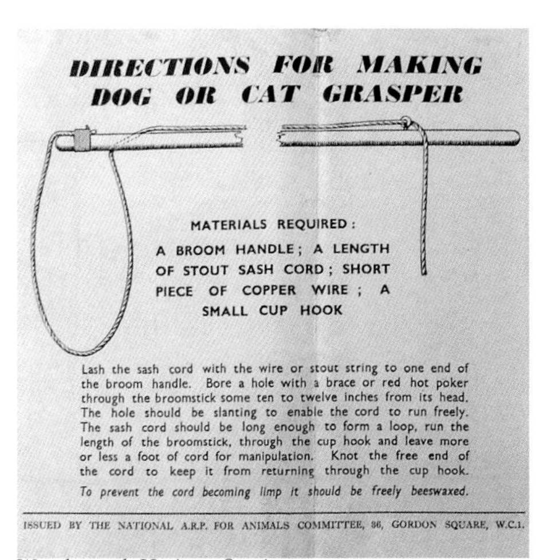

DIRECTIONS FOR MAKING DOG OR CAT GRASPER

MATERIALS REQUIRED:

A BROOM HANDLE; A LENGTH OF STOUT SASH CORD; SHORT PIECE OF COPPER WIRE; A SMALL CUP HOOK

Lash the sash cord with the wire or stout string to one end of the broom handle. Bore a hole with a brace or red hot poker through the broomstick some ten to twelve inches from its head. The hole should be slanting to enable the cord to run freely. The sash cord should be long enough to form a loop, run the length of the broomstick, through the cup hook and leave more or less a foot of cord for manipulation. Knot the free end of the cord to keep it from returning through the cup hook. *To prevent the cord becoming limp it should be freely beeswaxed.*

ISSUED BY THE NATIONAL A.R.P. FOR ANIMALS COMMITTEE, 36, GORDON SQUARE, W.C.1.

Wandsworth Heritage Service

Imperial War Museum

State Library of Victoria

"WE'LL SAY IT'S WORTH BARKING FOR!"

Dogs of all breeds love and thrive on "Chappie". This makes it all the more embarrassing for us to warn you that this complete food is in short supply and that its sale, in all fairness, is restricted to old customers.

"Chappie" is the complete, scientifically balanced, all-round diet for dogs. Vets, breeders and other experts agree that it provides the essential nourishment for the promotion of robust health. If your dog is deprived of "Chappie", just tell him how sorry we are. Give him this message from his more fortunate brothers: "Cheer up, old chap, we know the good things you are missing. Bark for the downfall of Hitler. Then, when peace comes, see that your master puts you on 'Chappie'. We'll say it's worth barking for!"

British Library (Newspapers)

NARPAC carried on – while behind the scenes its constituent charities rowed incessantly. The Government despaired. Meanwhile not many of the public seemed to know the strange organization existed, but where they did, they were enthusiastic supporters. Above, pet-loving ladies queue to have their cats and dogs registered at an Animal Guard Post near the British Museum.

It was food, not bombs, however that would really dictate the fate of wartime pets. There was no food ration for companion animals and from mid-1940 it was a punishable offence to give them anything judged fit for human consumption. Manufactured pet-food like Chappie (*above*) made of horsemeat was restricted before disappearing completely.

pets. 'It is the pampered and well-fed dogs of the better-off classes that make inroads into feeding stuffs that might be better used,' he was told. How true that was. There was trouble ahead.

Concern was noted meanwhile in *The Animal World* among tortoise owners about the high price of lettuces. It was suggested they be given 'bananas or oranges' but such luxuries were by now beyond reach: 'Some will take pieces of apple but many do not seem to appreciate this.'

On 7–9 April 1940 the Germans overran Denmark. Danish bacon and butter was now off the British housewife's menu. German naval and air landings rapidly seized southern Norway. British and French troops hung on perilously in the north. On 7 May, 'the Norway debate' in the House of Commons made it clear that Neville Chamberlain, moth hunter and bird lover, must go.

On 10 May the Germans invaded Belgium and the Netherlands. Another food larder had fallen. Pet fanciers had their own view of events. When Holland fell *The Aquarist* magazine regretfully recorded: 'Away goes our last source of imported stock.' It had been the origin of such fancied fish as golden orfe, golden rudd and golden carp.

'Our hobby,' said the journal, 'can give a peace obtained by few' – and was the only pet hobby which could be kept going in an airtight room.

Winston Churchill formed a coalition government in London. By the 13th the Germans were across the River Meuse. The next day the new home defence force was announced for men between the ages of 15–65, to be called the Local Defence Volunteers. A quarter of a million offered their services within twenty-four hours. There was nothing phoney any more about this war, for humans or for pets.

There was a war already raging in Downing Street. 'Bob' and 'Heather', the Dover-sole-rewarded Treasury felines of the appeasement years, were still in charge. Churchill's youngest daughter, Mary, recalled the fate of the incumbent(s) when her father took over: 'He was treated with great kindness but we disrespectfully named him Munich Mouser since he was a holdover from Neville Chamberlain.' In fact there seem to have been two appeasement era cats.

Mr Churchill meanwhile had formed a special relationship with the Admiralty cat he had encountered on his return to the Government eight months before. He had named him 'Nelson'. Already he was a bit of a newspaper celebrity.

The Washington Post's London correspondent sensed the tension at the heart of power. 'Nelson will follow his master shortly to Downing Street and make a problem of protocol. How, it is asked, will the Munich cat react to Nelson? Will he follow Chamberlain next door to his new home at No. 11 leaving the field at No. 10 to Nelson? Or will he refuse to abdicate and call for a show-down?'

It was the former. Lady Mary Soames recalled that Nelson chased the *ancien régime* cat (or cats) out of Downing Street pretty sharpish, though maybe not entirely. Secret Government files reveal that Bob 'became the pet of Downing Street staff' when the Treasury was hit by a bomb in late 1940 and Heather lost a fight with a rat in a Treasury storeroom in 1941. Bob seems to have survived skulking round the garden entrance while Nelson took up official residence.

The Cabinet Office in nearby Great George Street had acquired a rival cat, 'Jumbo', sometime in 1939, who had an official weekly maintenance allowance of 1*s*. 6*d*. for food under the name 'Mr. J. Umbo'. Jumbo's death certificate, dated 8 June 1942, signed by the ODFL vet at the Eccleston Street Animals' Hospital, is preserved in the National Archives.[14]

14 Cats seem to have been at the heart of Britain's war leadership, in contrast to Adolf Hitler's doggy entourage. According to the biographer of his mistress, Eva Braun, the Führer 'was allergic to cats'. And look how it all turned out.

Chapter 10
The Dunkirk Dogs

The Belgian Army surrendered on 28 May 1940. Across the Channel, the British Expeditionary Force began its long retreat to an embattled pocket on the northeast French coast. An extraordinary episode in the story of wartime domestic animals was about to unfold.

Throughout history, armies have always attracted dogs. There's something about field kitchens perhaps, or general, barking excitement. And young men respond with equal puppy-dog affection, compounded by loneliness and a longing for an affectionate companion. It was the same with the British Expeditionary Force, especially in retreat. Second Lieutenant E. J. Haywood was a young infantry officer. He recalled the mournful slouch through Belgium in May 1940:

> We passed through village after village. Dogs were whining and running about, vainly looking for their owners. Other dogs had been left tied up, and barked furiously at us, or howled dismally.
>
> To my great annoyance, a contingent of dogs of all shapes and sizes decided to join the procession. These wretched dogs were obviously strays from Bambecque,

> and were evidently prepared to forget their newly acquired problems of food and shelter in the transient joy of a walk that had more canine flavour than usual to enliven it.
>
> I cursed them all bitterly and fruitlessly, for, as they yapped and frisked about, they advertised our approach.

Edward Oates, a Royal Engineer, remembered shooting the dogs: 'The dogs were just roaming about, there was nobody there, and they were getting dangerous, you see. They were getting hungry and snapping a bit.'

Sapper E. V. B. Williams (an RSPCA staff member now serving in the Royal Engineers) took a kindlier approach. He recalled seeing the 'pitiful procession of refugees, pushing handcarts piled high with personal belongings and sitting on top a small dog quite content and taking stock of its new surroundings. But in some way or another it never took us long to make friends with these gallant pets. Also in quite a number of cases we were able to give these animals a painless end. This was the ultimate fate of our little camp follower, "Sapper". She was a small black and tan terrier who wandered into our camp the day we had landed in France. She was about seven weeks old when she joined us and was with us the whole time until things got too hot.'

Poor Sapper did not make it to the beaches – it appeared she got a British bullet between the eyes. But there were plenty of pets who did make it. The RSPCA reported: 'Large numbers of dogs were gathered with the retreating British troops in the Dunkirk area.'

As Royal Navy destroyers came alongside and the 'little ships' took bedraggled men out to the bigger ships offshore, amazingly, animals were scrambling into the sea and onto the rescuing ships – 'knowing with the sureness of canine instinct that the men who had so-far befriended them in their appalling need would not desert them at the last,' as the RSPCA's post-war history put it.

The story would be told of 'Boxer', a brindled Bulldog who was taken *to* France by his owner, Capt. C. Payton-Smyth of the Royal Army Service Corps, when the BEF shipped out. In the débâcle, Boxer, who hated water, became separated from his master and had to be physically thrown onto a rescuing ship. They were later happily reunited.

Ordinary Seaman Stanley Allen, then aged twenty, aboard the destroyer HMS *Windsor*, recalled: 'There was a little dog, a terrier-type mongrel, which came on board with some of the soldiers. He only understood French. When I spoke to him he wouldn't leave me. That little dog came back with us on the other two trips that we made.' He called the dog 'Kirk'.

Seaman Norman Battersby told his son years later that he gone into the blazing port with a Royal Navy shore party and 'evacuated a very special person' – a stray cat that he found hiding among the rubble and wreckage. He tucked the cat under his coat and smuggled it home, where it lived contentedly, a much-loved family pet, for several years to come.

Our Dumb Friends' League reported 'one of the little ships put out to sea from the quay at Dunkirk leaving behind an Alsatian, which had been unable to get on board. The dog dived into the sea and swam for its life. A British naval rating dived into the sea and brought the dog to safety. It had been quarantined by the League. The dog's owner has vanished but its rescuer will take it when he can.'

'Men of the French Army rescued from Dunkirk and Calais had been accompanied by twenty-two dogs and one cat,' said the reports. Two were snatched from German boats captured by the Royal Navy and one from a Danish ship. ('The German and Danish dogs have now been placed in English homes,' the League would later report.) The *Daily Telegraph* told the story of a French soldier with 'a large, fierce looking Alsatian,' coming ashore at an English port. 'I found him in Lille. He marched and fought with us all the way to Dunkirk. I bring him over on the boat; he is staying, I think,' said the Frenchman. And, according the *Telegraph*, the Alsatian did just that.

Meanwhile, *The Cat* magazine got cross about pictures in the press of 'returned soldiers with Belgian kittens'. Did people not understand that without paying for months of quarantine these poor creatures must simply be destroyed? Better they were allowed to stay alive in Belgium whatever the conditions were like there.

Thus it was that hundreds of cats and dogs were involuntary participants in the 'Miracle of Dunkirk'. Amid the chaos there must be order. What about rabies? What about quarantine? The RSPCA recorded:

> All along the Thames Estuary and the south and southeast coast where the little ships of all kinds landed wearied men, equally wearied and homeless dogs were also brought ashore. The danger to this country's animals resulting from an outbreak of rabies in such circumstances is too obvious to be stressed and R.S.P.C.A. Inspectors did a great work by caring for these animals until something could be done for them.
>
> Frequently the soldiers would leave the ships and the dogs would still be aboard. In such cases the Society's Inspectors would have to round them up.

The Folkestone inspector found himself in the heat of battle when the ship he was on returned to the beaches still with a cargo of dogs – to pick up more men and their pets.

Kirk, the HMS *Windsor* dog, made it. His rescuer would record later: 'After Dunkirk was all over, he was collected by a PDSA van to go into quarantine for six months before he was taken on to the staff of the parish where our Sub-Lieutenant's father was vicar. All of us cheered the old dog off.' But many were not so fortunate: 'Where the owners could be traced, the RSPCA placed a few dogs in quarantine at its own expense but the majority of the animals had to be painlessly put down as they were either severely wounded or suffering from hysteria owing to the strain they had undergone.' They were innocent victims of a conflict they had not sought and could not possibly understand.

And among the animals (but these were definitely not pets) that did not get away were some 1,700 Indian Army mules of so-called 'Force K6' sent to France to carry ammunition and stores for the British Expeditionary Force (BEF). One company was captured outright in the German advance, three more (each 400-strong) were left behind at Dunkirk and St Nazaire.[15] The men got away.

As news of the Dunkirk pets' plight got out (the animal welfare charities could not resist the publicity), 'there were many inquiries from animal lovers anxious to know whether arrangements can be made for individual soldiers to keep these animals,' reported *The Animal World* in the July. The RSPCA was unsentimental: 'All dogs entering this country must undergo six months quarantine at an approved kennel,' it sternly stated. 'The cost of maintenance for this period is around thirteen guineas.' And most of them were 'camp-follower dogs, not attached to any

15 From the war diaries the animals appear to have been simply set free or with 'great reluctance handed to the French'.

individual,' so *The Animal World* reported. Their fate looked bleak:

> Hard facts cannot be ignored. There is a grave risk of rabies and full quarantine would entail heavy expenditure. Where it is not practicable or desirable to keep these refugee dogs they will be given a speedy and humane end. At one port, 66 have already been put down, 20 at another.

The French Army dogs had 'been segregated, provided with leads and given exercise and are being looked after until their owners go back'. The fate of the lone French military cat that escaped the débâcle was not recorded. Presumably it was not there to answer General Charles de Gaulle's call for Free France to fight on.

Our Dumb Friends' League was less quick to condemn. 'Many dogs attached themselves to the retreating British armies,' stated its 1940 report. 'What was to happen to these dogs? The answer was destruction or outside help. When asked by the Ministry of Agriculture for help, the League came back at once. We will take them! Send them to us at once! So the League has or had 107 dogs in quarantine.'

There were ninety-five dogs, and one cat who had fled for safety with Allied civilians (rather than the military), so the League reported. Dumb Friends' had 'adopted the same policy of paying for their quarantine' and although this had attracted anonymous criticism, 'there was not just a national, but an international duty to save them.' Formal thanks had been received from the Belgian and Dutch Governments in exile for the League's humane intervention.

There was 'not a little difficulty caused by the separation of dogs and cats from their refugee owners,' the RSPCA reported more sternly, resulting in 'many inquiries made of quarantine kennels. In many cases the animals were traced but there are a few sad hearts among French or Belgian people who have lost trace of their canine or feline friends' – often the only remaining link with their former homes.

It would also seem that there were pet rescues by air. *PDSA News* would later tell the story of 'Fortnum', a 'French red squirrel who flew to England with an RAF Flight-Lieutenant in May 1940' and who later made 'more than half a dozen trips over enemy territory'. 'The squirrel was handed over to Miss Phyllis Kelway, a naturalist author and now lives in a hutch of rabbits,' it was reported. What a story he could tell!

The RSPCA's 1940 report gave an account of 'an RAF Squadron Leader's dog – lost in the difficult days before returning home, the animal was last seen somewhere in northern France'. The dog, however, had had a collar plus a label: 'Inquiries were set on foot although it seemed doubtful if the dog could have reached England unaided, but in the end she was found being cared for in a south coast kennel. Somehow, no one knows by what route or vessel, Nora had come over alone, guided by her own instinct and sagacity'. It looked very much as if 'Nora', the air force dog, had been flown out.

In the chaos, plenty more dogs simply slipped though. 'Many dogs were smuggled by the troops from the ports far inland, but common sense eventually prevailed and these dogs, one of which travelled as far as Leicester before being detected, were placed in quarantine and eventually found good homes. Those hopelessly injured or diseased were painlessly destroyed,' said the RSPCA's journal. It expressed particular pride in 'that it was enabled to bring over the last dog from Dunkirk'.

The National Canine Defence League announced that it

too was paying for quarantine for French and Belgian refugees' dogs, 'just as it had done for those from Austria and Czechoslovakia'. 'The Dogs from Dunkirk are the latest canine sufferers from Nazi terrorism whom the League has helped,' recorded *The Dogs Bulletin*.

The League was remarkably upbeat about the war's progress. 'The BEF dogs from Dunkirk are now in quarantine,' said the *Bulletin*. 'They will doubtless emerge in six months time to find Britain at peace.' Or invaded and conquered.

For Britain's pets the real time of trial was about to begin.

Chapter 11
Three Million Dogs to Die

So far no German bomb had fallen on the nation's pets. Some people were asking what was NARPAC for? On 10 June 1940, Edward Bridges Webb (who had just been made full-time controller of the organization's registration branch) wrote to the Ministry of Home Security to say that 'his' Technical Officers 'felt they had nothing to do of national value, standing by lest the public be attacked by panic stricken animals in the case of air raids'. There had been lots of panic but not yet on the part of pets.

A morale-raising letter from Sir John Anderson or a placed parliamentary answer would boost drooping morale, he said. The minister had to be reminded what NARPAC was. It was a voluntary organization covering the whole country, not just towns, but now farm animals as well with the recent launch of the so-called 'rural scheme'.

Statements of purpose appeared in the press. NARPAC was there, both 'to protect the public from injured, frenzied and gas-contaminated animals at large' and to 'advise livestock owners on killing seriously injured animals and salvaging their carcasses for food purposes'. Farm animals were not the immediate problem. It was domestic animals in the seaside towns of southern England where the bedraggled survivors of Dunkirk were still coming ashore which were now in the front line of a German invasion. Like the inland cities before, a new effort began to empty them of children. And children had pets.

It was announced publicly on 30 May: 'The following towns on the southeast coast are to be declared evacuation areas: Great Yarmouth, Lowestoft, Felixstowe, Harwich, Clacton, Frinton and Walton, Southend, Margate, Ramsgate, Broadstairs, Sandwich, Dover, Deal and Folkestone.

'Arrangements are being made for those children whose parents wish them to go, to be sent from these areas to safer districts in the Midlands [some would travel as far as the Peak District] and Wales. The movement will start by special trains leaving next Sunday.'

The RSPCA moved in to rescue the seaside school pets, as it had in London nine months before. There were tearful separations all over again. From 13–18 June 1940, around 100,000 children were evacuated (in many cases *re*-evacuated) from 'invasion corner'. As many parents and children were getting ready to follow. Would there be another wave of mass destruction of cats and dogs?

As part of the new firmness of purpose, on 26 May prominent members of the British Union of Fascists were interned under wartime emergency powers. The BUF had political and personal links with anti-vivisection activists. Our Dumb Friends' League had arranged with Scotland Yard to take their pets into care. Apparently included was 'Goldie' the 'beloved ginger cat' of BUF leader Sir Oswald Mosley.

That same month came a general round-up of enemy aliens, Germans and Austrians, many of them Jewish refugees. Italians were also included, even though Britain was not at war with Italy until 10 June – all to be sent to improvised barbed-wire-strung camps amid general spy-frenzy. The fate of their pets was nasty, brutish and short. On 1 July over 1,000 internees were drowned when the liner shipping them to Canada was torpedoed. Among them was an Italian called 'Sig. Azaria'. An 18-year-old Londoner, Colin Perry, had known him (he had clipped the family canary's claws). He wrote in his diary:

> The man who owned the pet stores at the bottom of our road has lost his life in the *Arandora Star*. 'The pet man', we called him, it seems incredible that an old man who kept a pet shop in Upper Tooting Road had suddenly been snatched away to forfeit his life in the Atlantic.

And what had happened to the Tooting pets?

The German animal welfare magazine, *Reichtier-schutzblatt*, meanwhile showed the triumph of the Blitzkrieg with photos of Hitler, the 'animals' true friend', patting horses on the forehead and reaching out to pet 'a French battlefield stray'. The little dog looked quite unperturbed by the Führer's caress.

The PDSA's animal hospital at Bièvre, southwest of Paris, was overwhelmed, its staff lethalling a tide of refugee pets and exhausted military animals before its English staff members got out by car and on foot at the very last gasp. The clinic, endowed by an American millionaire, became a hospital for German Army horses.

This really was 'backs-against-the-wall'. In London a secret propaganda committee 'in a position to tap into and guide every source of public expression' agreed on 10 June to fire up an 'anger campaign' to rouse the people against the 'fundamental rottenness of the Germans, the secret beast in the middle of otherwise civilised Europe who kills without pity as animals kill. They know no law as animals know no law.' The weakest link in the chain of public courage is the lonely woman, the propagandists agreed, who could be moved to righteous anger by tales of enemy cruelty to the vulnerable.

Results were swift in coming. A story reported 'in neutral newspapers' was picked up by the London press on the 13th. 'Three Million Dogs to Die in Germany' it stated starkly, 'because, according to an official explanation, they need food that might be used for human consumption.' How rotten, how beastly! Actually it was politically very useful considering what was being discreetly discussed in Whitehall.

Animal lovers were appalled when they read it. Martina Corfe, aged 26 and an auxiliary ambulance driver in Greenwich, reported in her Mass-Observation diary that her friend, Mabel Cooper, 'was more upset than anything else by the slaughter of all the dogs in Germany, she keeps talking about and crooning affectionately over her own dog'.

Edgware schoolgirl Helena Datz saw 'a newsboy [who] had drawn a crowd with his completely unintelligible shouts. The cause of the furore was the command to kill all the dogs in Germany. My friend made a feeble joke about that including Hitler and all the Nazis.'

It was in the papers, on the BBC News. The American press would pick it up. 'Can it be true? Three million dogs doomed to death in Germany?' asked *Our Dogs* on 21 June. 'This hideous holocaust must be the greatest slaughter in the history of dogs,' wrote columnist Thelma Gray, Corgi pioneer and royal confidante. She had an inside view when she wrote:

> From a dog breeder's point of view it is a disaster of the first water. Without German stock, quality will

> diminish. Alsatians have varied blood line and as for Wire-haired Dachshunds, Schnauzers, and Boxers – we could get by with stock from Switzerland and America – but what of Rottweilers?
>
> Alas the poor rotties, there is not enough material here and America cannot help. Germany has all the best stock and they are condemned to death. What a tragedy for a fine breed fast becoming better known in England.

She was sure however that at least four dogs would survive – the two Chows which she had personally sold to Reichsmarschall Göring and the two Corgis belonging to Herr von Ribbentrop. 'These two Nazi officials are very fond of their dogs and I can't help thinking that they will wangle things so that their pets get off unharmed,' she wrote.

The humane Amy Golightly wrote in 'Kennelmaid's Corner': 'It must have been with horror that each of us heard the broadcast telling us that over three million dogs in Germany are to be destroyed. No matter what else they may be, the majority of these owners are fond of their dogs.

'They have shared their meagre rations with them and they have tried, as we under like circumstances have tried, to spare the horrors of war from them,' she wrote. 'To know that, if we are careful and sensible, such an edict is unlikely ever to go forth in England, must be of great comfort to us. There Must Be No Waste!' Actually, she was on exactly the right track.

A Home Office official made a note about the German 'dog destruction order'. He saw it as a bid to get the 'glycerine and the fertilisers which their destruction by scientific method would provide'. With the 'national interest in mind,' he suggested, could not the same be done in the south coast towns being evacuated? 'I gather that Morel's [a firm of animal renderers] are in Folkestone collecting dead dogs and treating them already.'

While invasion panic gripped the seaside, a new scare stalked the wider nation. It had begun with the German move into Holland when paratroops had seized bridges and airfields. The first admission that Britain might be readying a canine answer was a statement in Parliament on 21 May that 'training experiments [were] now in progress' with patrol dogs.

Britain's tentative war-dog trials were indeed in progress, in the hands of the Crufts 'grandmaster', Mr Herbert Lloyd.[16]

The War Office had agreed in late April 1940 that four dogs should be trained at Mr Lloyd's kennels at Washwater, near Newbury (they would soon move to the requisitioned nearby Woolton House), with a junior officer, Lt. H. A. Buxton, and six other ranks as the beginnings of an elite dog-squad. The Labrador-crosses were set to go to France with the BEF on night patrols, but the Dunkirk débâcle had ended that plan.

Alsatian enthusiast Major James Y. Baldwin (whose doggy impact on the war would be considerable, *see also* p.231–2) expressed his anguish in early May in *The Dog World* – in an article condemning the 'British government for saying no to dogs which can help to win the war.' An old War Office chum dismissed him as a nostalgist: 'But

16 In his 1938–39 experiments for the Home Office, the famed Cocker Spaniel breeder had defined the difference between a 'dog', which 'worked for the sheer love of pleasing his handler', and a 'hound', which worked 'only to please himself'. A hound 'would give up if distracted,' he said, whereas there was 'no limit to a dog's intelligence and desire to please.' According to him the most 'pliable and humanised dogs' were Labradors.

you see, my dear chap, how different things are from what they were last time.'

Alsatian owners who had offered their dogs for war service had been told to enlist as police Special Constables. A proper start on mobilizing the nation's dogs must be made now, said the Colonel.

The old canine warrior, Lt-Col E. H. Richardson, was also anxious to get back in the fight. An Airedale man, he had run the Royal Engineer dog-training school at Shoeburyness from 1916–18. Airedales, Lurchers and Collies were splendid under fire, but Hounds were 'useless' and Poodles too playful, he had found. His book, *British War Dogs* (1920), had been a canine bestseller. It concluded: 'A dog trained to do definite work is happier than the average loafing dog no matter how kindly the latter may be treated.'

Maybe so, but some dogs preferred loafing. But this was war and lazy dogs should now be hunting Nazi parachutists. From before the outbreak of war, the Colonel had been experimenting with gas masks for gallant message-carrying dogs at his famous kennels 'Grassland' at Horsell Common, Woking. His bizarre-seeming freelance efforts had been widely publicized in the press.

In the June crisis, Colonel Richardson sent a telegram to the War Office: 'When war broke out, I urgently asked that I might be allowed to collect and organise the dog power of this country [but was ignored]. Dogs are invaluable for defensive guarding all vulnerable, secret and dangerous places. Apologise for wiring but can get no attention paid.' Now was the time for canine action.

The Berlin dog paper, *Die Hundewelt*, reported on a grand recruiting rally that had added 15,993 Airedales, Boxers, German Shepherds and Doberman Pinschers to the Wehrmacht's ranks. Stories of the German rations-only-for-war-dogs decree appeared in a scandalized British press. Did it just not show an absolute determination of purpose to win the war?

There were questions in Parliament. The Secretary of State for War had been asked in May 'whether he was aware that the Germans had something like 100,000 dogs', all seemingly ready to pounce. Mr Anthony Eden referred to 'training experiments now in progress' with the British Army. At the time, although he did not admit it, there were just four official war dogs.

Clearly there should be many more for there was no shortage of talent, human or canine. A search began for dog handlers already serving with HM Armed Forces.

The Dog World picked up the doggy rumours, referring to the 'experimental work being done by Mr H. S. Lloyd'. 'One can easily imagine an Alsatian dashing forwards and attacking a parachutist,' said the ever-eager journal. Just look how clever dogs had been in pre-war dog show intelligence tests. 'One could surely coach one of these intelligent little fellows to spot one of these "hell's angels" in whatever guise they might appear.'

A correspondent for *Our Dogs* would comment: 'In my opinion, the people best suited to guard us against the possibility of surprise landings are those men and women who know the country well [and] can ride across it. Shoot straight and command a following of powerful well trained and loyal dogs.'

The Times that summer visited Yorkshire 'hunting country' to find the 'last remaining remount depot' for British Army cavalry horses. It all seemed a bit archaic. 'There are still several hundred horses in the establishment,' their correspondent noted. 'There must be many areas where local hunts might form the nucleus of a mounted Home Guard. There will probably be little fox hunting this year, the hunt for parachutists could be a substitute.' Fox hunting might yet find a patriotic role.

Animals generally had a place in the invasion defences. A veteran wildfowler pointed out in *Country Life*: 'The green plover for many months of the year will give away the presence of any intruder in the marshes.' Raucous herring gulls would apparently do the same in rocks and dunes.

'Dogs and the Invader,' an NCDL pamphlet advised. 'Guard dogs should be kept inside to give early warning of the approach of a parachutist, kept outside on a leash they will be easy target for enemy tommy guns.'

On 20 June the first German air operations began over Britain. The still-voluntary flight of women and children from invasion corner turned into a flood. Just as in the panic-stricken late summer before, planning went ahead on the basis that 'no cats or dogs may accompany owners [should they be] compulsorily evacuated'. It was clear that a tide of refugee pets would not be welcome at 'the reception end'. 'Poultry and rabbits should be disposed of by their owners before evacuation,' the Ministry of Agriculture advised. Actually, a nice rabbit for the pot might be quite welcome.

Three days later the PDSA held its annual Animal Lovers' Parade and service at St Martin's in the Fields, Trafalgar Square. Bowler-hatted, tweed-jacketed, riding-breeched Technical Officers marched in formation behind their banner with a strong female contingent and plenty of pets. The Dispensary's motor ambulances rolled through the streets, polished and gleaming. They had yet to see action under fire; it would not be long.

On 29 June the War Cabinet Home Policy Committee approved draft legislation as an Order in Council – 'for animal control in the event of hostile attack' – delegating slaughter powers to NARPAC 'mobile units' in the evacuation zones. It could be their finest hour. The Committee would devise a provisional service for the humane destruction of cats and dogs in the towns concerned with a fleet of small ambulances specially equipped to do so.

It was all terribly thorough: railways were consulted about their staff taking over animals brought to the stations by evacuees. They would not be allowed on trains chuffing to safety. 'A room for cats and a yard for dogs should be provided until they can be disposed of,' Sir William Wood, vice president of the LMS Railway, was informed. 'It would be most helpful if some empty boxes or baskets with lids could be produced for the temporary reception of cats.'

In the face of this new crisis, the supposed solidarity of the animal welfare charities was already breaking down. Mr Frederick Donne, a vet from Westcliff-on-Sea, told the *Veterinary Record* that summer that he had been an enthusiastic organizer for NARPAC since its inception, 'personally enlisting 255 Animal Guards, registering many thousands of animals and collecting £85s. 11d'.

Now with the prospect of evacuation: 'Notices are repeatedly being posted up with reference to same. The latest poster is the RSPCA's adorned with the pictures of the head of a dog and of a cat. This poster has caused much alarm and distress to owners of pets.' It had the chief constable's name appended so looked terribly official and read thus:

> PLEASE CONSIDER THIS. If evacuation of this town is ordered, it may not be possible for you to take your pet animals in the train.
>
> Animals will be painlessly destroyed by competent persons at the following addresses …

It mentioned two animal societies, two corporation depots and a vet in Southend, but not Donne. 'I have put up a huge notice on my door,' he wrote. 'Notwithstanding the poster, this veterinary surgery is remaining open.'

The poster had been put up on display by a Mr Stephens, the RSPCA's Inspector in Southend. He could inform his headquarters meanwhile that there was no danger of wild animals prowling the promenade as he personally had destroyed the lions of Bostock's Menagerie at the Kursaal funfair on the outbreak of war.

Superintendent F. C. Rogers, a PDSA Technical Officer in Southend, later wrote: 'The worst time was after the Dunkirk evacuation. The scenes I witnessed were beyond imagination, please God they will never be witnessed by anyone again. That morning a notice had gone up that the enemy were about to invade with a panic evacuation as a result. To the PDSA fell the sad but merciful task of putting the animals to sleep. It was heartbreaking – all of them, the animals, had been enjoying life to the full and so jolly too.'

*

The slow-footed NARPAC had let the charities go in and act on their own. The Committee must fight back. On 30 July Colonel Stordy told the Ministry of Home Security that he had written to 'veterinary officers, chief constables and town clerks of all the towns in the coastal defence district from Northumberland to the Dorset border and ten miles inland' to tell them to expect the wide-scale destruction of cats and dogs. NARPAC would undertake the humane destruction of pets but the digging of trenches for the carcasses should be the responsibility of local authorities.

It was a bureaucratic disaster. Except in the original

designated nineteen coastal towns in the southeast and East Anglia, the population of everywhere else was supposed to stay put. NARPAC had gone completely against policy and caused an almighty flap. The circular must be withdrawn and a shamefaced Stordy duly did so.

The general panic about evacuation, invasion and pets was heightened by the tragedy unfolding in the Channel Islands. In the weeks between Dunkirk and the arrival of the Germans on 2 July, there was a mass scramble to get out. Prize cattle herds were put on boats for England. Evacuees could only take what they were able to carry. Domestic animals were simply abandoned. Parish constables rounded up pets from empty houses to be put down. 'As there was no room on the boats for pets, the dogs and cats were shot and so great was the run on the vet's services that the owners had to line up while their dumb friends were dispatched,' said a press report.

Betty Hervey was a six-year-old child on Guernsey. Sixty years later she recalled:

> All our animals had been turned loose to fend for themselves – my black lab, some cats, chickens and rabbits. At the harbour everyone was weeping. Older children were leaving without their parents. We crossed the channel in a cargo boat, and there were cattle in the hold.

The 'down-with-all-dogs' clamour got louder. A branch of the National Farmers Union passed a resolution that the canine population be reduced 'except for useful working dogs'. On 5 July in *Our Dogs* doggy author M. Douglas Gordon raised the standard against 'the squealings of the anti-doggites' and their attacks on 'hounds, pigeons, rabbits etc. as useless consumers of valuable food'.

'We must bestir ourselves to contradict such stuff,' declared Gordon. And on the alleged Nazi canine holocaust, he wrote: 'It is by no means certain that Germany has destroyed all her dogs. Later reports suggest it is only pet dogs that have been slaughtered, which puts a very different complexion on the matter.' The Canine Defence League also roared back mightily:

> Fears are expressed that our dogs may have to be sacrificed in order to help us win the war and we find that the dog haters are trying to take advantage of the national danger to see our friends exterminated.
>
> The monomaniacs who detest dogs are anxious to seize any chance that our extremity may present in order to gratify their splenetic and venomous desires. It is folly to imagine that such people do not exist and that there is no clash of interest in the food supply or the [shipping] tonnage question.

It was 'isolationist folly,' said the League, for pedigree dog breeders to call for mongrels to be eliminated and only show-dogs allowed to live. Dog lovers unite!

By now the 'three million-dogs-to-die' story had reached America. Was this propaganda yarn going to rebound? *Popular Dogs* magazine made inquiries, to receive at last a reply from the Doberman Pinscher Club of Germany: 'We rejoice that American Doberman friends take such kindly attitude towards our breed but most of our dogs are at present with the Army,' its chairman had said, '[but] as long as this war, forced upon us, remains undecided, even the best dogs should and must be called into service.'

Nazi dogs were top dogs indeed.

Chapter 12
No Cat Owner Need Worry

Summer 1940 was a tough time to be a beast of England. Hundreds of seaside pets were going into the lethal chamber. Urban dogs were biscuit-less, cats were the enemy within. The Germans were busily making their invasion plans (which included the employment of 60,000 horses) and black-crossed aircraft prowled the Channel, testing the mettle of RAF Fighter Command.

Newspapers broke the truly awful news on 6 August: 'Penal Offence to Waste Food: Warning as to Meat for Dogs' ran the headlines. It could not have been more ominous:

> The Ministry of Food to-day made an order coming into force next Monday for preventing waste of food and carrying with it penalties against offenders. It will be an offence to waste food, which is described as everything used by man for food or drink other than water.

It was further explained: 'Under the order it is not forbidden to give meat to dogs, but anyone giving an excessive amount would be liable to prosecution.'

But who might judge what that meant? The announcement was personified in Mr Robert Boothby, MP, Parliamentary Secretary to the Ministry of Food, who reassured pet owners that 'it was more in the nature of a warning', and he did not expect a lot of prosecutions. The penalty was two years' imprisonment or a £500 fine, or both. Boothby played down the prospect that 'inspectors would be sent to houses to see if the regulations were being carried out'.

The result was utter confusion. He had declared at a press conference that it was indeed an offence to give food to an animal that could be consumed by a human being. What about giving stale bread to a pet dog or a saucer of milk to a cat? The Ministry was besieged with questions.

There was more controversy when the editor of *Cage Birds* gave a talk on the BBC which seemed to recommend the feeding of foods suitable for human consumption to poultry. If they could not get birdseed, then it was bread or oatmeal he had recommended cheerfully but doing so was now illegal. The RSPCA advised on seed-producing plants that could be grown in 'sheltered sunny positions' including linseed, millet, sunflower and hemp – *Cannabis sativa*.

Evidently there were other ways to keep the family budgie fed. A Nottingham housewife confessed to the social historian Norman Longmate after a gap of thirty years having falsely declared to a shopkeeper that she had two birds to get her budgie, 'Mickey', a bit more. Such birdseed as was available was often sold without millet, its most useful ingredient. A Lancashire housewife remembered millet costing £1 per pound on the black market. A Rotherham woman recalled even ordinary birdseed reached this price but was so un-nutritious that the family budgie simply starved to death.

The Minister asked his advisers what had happened in the last war. In 1918 a number of prosecutions took place of people 'wasting human food by giving it to dogs' in defiance of regulations, he was informed. Unpatriotic pet owners giving bread and milk to Pekingese and meat to St Bernards were two cases in point. This was clearly a political minefield.

The RSPCA waded in. It pointed out that the Waste of Food Order did not mention animals at all. 'It does not seem possible that any court would hold that the moderate feeding of any pet upon food suitable for human beings would be a waste. Do not deny pets your table scraps because some well meaning person tells you you are liable to a penalty,' it announced.

The Cats Protection League did not know how to reply to a flood of anxious letters – 'the matter was taken up with the Ministry of Food, from whom a very sympathetic reply was received' so it was reported in *The Cat*. Prosecutions would only be cases where, 'human food has been used wastefully, either because the quantity was excessive or because other food stuffs were available,' readers were assured.

Meanwhile cats had a card up their sleeves: mice. As *The Cat* asked, 'is it antipatriotic to deprive the community of so many foodstuffs for one's own pleasure? If the cat were merely a pet, giving no service in return for his keep, the reply would be "Yes". But the cat is one of the best food protectors known. He saves the community literally millions of pounds worth of foodstuffs annually. Immense quantities are lost to rats and mice.

'If you live in the country the range of your cat's operations will extend far beyond your house and grounds,' insisted *The Cat*. 'Not only the mice that eat the household food, but rats, field mice, shrews, voles, etc., that attack crops and vegetables will be driven away or killed.

'No cat owner need worry about the rightness of keeping a cat in war time,' it concluded.

And since the drift of evacuees back to the cities, 'people have found that mice were overrunning their homes, eating their cereals and nestling in cupboards' so it was also reported. Following the unwarranted destruction of so many pets the year before, 'dealers now have waiting lists of people who wanted dogs and especially cats'.

But it was now formally illegal to give cats milk. The County Borough of Sunderland applied for cereal coupons to feed municipal swans. Was feeding the ducks in the park on breadcrumbs illegal? It might well be. The RSPCA pressed the Ministry for clarification.

The Duchess of Hamilton stirred with a long letter to *The Times* headed: 'Feeding Dogs and Cats' – 'The Council of the Animal Defence Society wish to remind owners that the order does not impose the destruction of animals,' she and Louise Lind-af-Hageby jointly declared. 'The order forbids the wastage of food. While the order contains no reference to the feeding of animals, Mr. Boothby has stated that it may be taken as a general rule that food unfit for human consumption can be given to household pets.' She saw the real enemy in plain sight:

> Those constituents of the National Farmers Union, who now clamour for the annihilation of household pets, can point to the obvious fact that dogs and cats must eat in order to live, but, in considering 'utility', there are mental and emotional values which must not be forgotten, even in time of war.
>
> It is reported – whether accurately or not we cannot say – that in Germany all dogs not useful for the

> purposes of war have been destroyed. This country is certainly not in a state requiring such ruthless measures.

'Dogs gave comfort to many a lonely wife of a serving soldier,' said the Duchess, while 'the elimination of the feline population would result in a food-destructive increase of rats and mice.' As for food shortages, 'many dogs are voluntary vegetarians,' she suggested, while 'many cats have inclinations to a non-meat diet.' She would not find many votes among pets for that one.

Her Grace's letter was published on 12 August – the day Hermann Göring declared the 'Eagle Attack', the all-out offensive on southern England to crush the Royal Air Force. That same day Mrs Anne Walters, a farmer's wife, wrote to *Farmers Weekly* to complain about huge flocks of sparrows eating corn in the fields – 'the people's food,' as she put it. 'My husband has a rook man,' she said, 'but the sparrows are worse. I should be glad to know what if anything is being done to prevent it.'

Not just sparrows were swarming in the air. The Germans had turned on the defenders' airfields to smash the RAF on the ground and win air superiority. Bombs and bullets were now falling in the fields of southern England – along with shot-down aircraft and parachuting aircrew. The etiquette of offering tea to bailed-out enemy pilots was debated.

Huge aerial battles raged over southern England. A few stray bombs fell in the north London suburbs on the 22nd. The next day *Farmers Weekly* reported on the battle for the nation's survival: 'Stock casualties in Nazi raids, first big test of Animal ARP services.' It had urgent news:

> Most bombs in the intense Nazi raids of the past two weeks have fallen in rural areas. Farm livestock has

> come off lightly. First aid had been given to a heifer valued at £40, which had undoubtedly saved its life.

The next day Fighter Command's most critical week began as 'savage' bombing aimed to knock out the airfields defending London. On the 25th a straying night bomber hit the City. Battle was also joined in the fields of Kent. As *Farmers Weekly* reported:

> There has been no falling off in milk yields. Uninjured cattle show no after effects. One dairy herd lost a cow killed outright but an hour later the owner found the rest of the herd lying quietly around the edge of the bomb craters. Most liable to take fright were horses – which would career madly about the fields at the sound of low-flying machines.

The only way to get compensation was to salvage what could be for human consumption, the journal advised. 'Dead beasts must be promptly bled – within 20 minutes of death, cut the throat and if possible disembowel it.'

NARPAC's rural scheme of flying squads of vets and butchers appeared to be working. 'Cattle and sheep had suffered because of their tendency to herd together,' it was reported. Horses and pigs sailed through. Hardly any farm buildings had been hit. All livestock casualties were in the open.

It was tough generally being a farmer. Nazi bombs caused thin milk, it was reported on 30 August. A farmer summoned under wartime regulations for selling milk with low butterfat escaped prosecution when a Land Girl 'gave evidence that the cows were nervy and jumpy after the first raid'. The case was dismissed.

Meanwhile the propaganda battle to ensure the survival

of the nation's pets raged just as fiercely. On 23 August the 'Dogs of Britain Fighter Fund' was launched. As *Our Dogs* proclaimed:

> Dogdom's direct contribution to the downfall of Germany has begun. You too can join the movement to provide Britain with another weapon in her plan to attack to defend.

The move was at the suggestion of Miss Veronica Tudor-Williams, famous breeder of Basenjis, the barkless African hunting dog. Mr C. Dowdewell, Airedale lover, contributed two guineas and Smoky Snelgrove Pekingese 10*s*. 6*d* in the first week, along with scores more patriotic dog lovers. A total of £5,000 was needed. The Kennel Club became the official sponsor.

Hungry dogs would need all the goodwill they could muster. Food was still the issue. The Bloodhound breeder, Lady Johnson-Ferguson, wife of a Scottish baronet, got into trouble in late August when she wrote to the Food Minister saying that she had special biscuits made of wholemeal flour for her dogs. Why couldn't everyone else? It seems she was trying to be helpful.

A furious internal memo pointed out this was a clear violation of the Milled Wheaten Substance (Restriction) Order: 'Lady Johnson-Ferguson is keener on keeping her dogs fed than winning the war.' Her dogs had to eat something, though.

Country dogs had a better time of it. It was a question of finding food. Newly-married Mrs Joyce Ixer was in charge of a Spitfire repair depot, part of the Supermarine complex around Southampton, while tending a kennel of her beloved Alsatians (she had been smitten as a small child) at home with her husband. As well as bombing and invasion scares, she recalled over seventy years later, the daily task of getting meat from slaughtered New Forest ponies to feed them – which was kept in a running stream to keep the flies off. Her dogs thrived

That summer, novelist and journalist George Orwell spent some happy time at his rented cottage at Wallington with its clucking hens and rows of onions. He wrote in his diary for 18–19 August: 'Two glorious days. No newspapers and no mention of the war. They were cutting the oats and we took Marx [the poodle] out both days to help course the rabbits, at which Marx showed unexpected speed. The whole thing took me straight back to my childhood, perhaps the last bit of that kind of life that I shall ever have.'

This was to prove the case for him and a lot of other pet lovers.

Chapter 13
Blitz Pets

On the morning of 4 September 1940, the War Cabinet Committee on Civil Defence met in London to consider: 'Item 1 – The Restriction of Consumption of Food by Cats and Dogs.'

Sir John Anderson was in the chair. Lord Woolton, the newly appointed Minister of Food, had produced a short paper concerning not just pets but also 'dog racing and the maintenance of zoos'. Protests from poultry keepers that their feedstuffs were being severely restricted were considered. Racing greyhounds were apparently consuming oceans of milk and mountains of rice too. What were these useless dogs doing to help the cause of national survival? But as Lord Woolton reminded everyone, any drastic steps to reduce the dog population would arouse 'intense public clamour just as taxation of dogs did in the last war'. Interfering with the people's pets was politically contentious.

The Committee agreed: 'There should be no action [yet] to reduce the dog and cat population generally.' But the fact that moves had already been made to reduce the use of human foodstuffs in the manufacture of dog biscuits to one half of pre-war levels should be stressed to 'critics' (meaning those grumbling farmers).

In Berlin that evening Hitler made an exultant speech at the Sportpalast. In retaliation for a pinprick RAF raid on Berlin, he declared London to be a target. Britain's cities would be erased and the last island of resistance extinguished. That same evening air raids were made on port cities – Bristol, Cardiff, Swansea, Liverpool and Newcastle. The 'Blitz' had begun and domestic animals were in the firing line as never before.

As for invasion, the Führer promised not to keep the British waiting too long. He was 'coming', so he promised the enraptured audience, dog-biscuit shortage in England or not. The crowd screamed: 'Hitler! Hitler! Hitler!'

That day GHQ Home Forces issued a briefing note: 'Germany is presently engaged in an attempt to gain superiority over our fighter force. If a firm measure can be gained over the next few days then a full scale invasion may be attempted over a broad front in an area between Shoreham to Southwold.'

The Chiefs of Staff met in Whitehall on 7 September and concluded the Germans had completed their preparations. For the next few days the moon, tide and weather were all favourable. Regional commissioners were teleprinted: 'You must assume invasion might be attempted at any time now though it is not assumed as a certainty. Civil departments to be so informed.'

At around tea-time that afternoon the first wave of bombers arrived over the capital, followed by a second, two hours later. The attack started in the East End before creeping westwards to central London, leaving more than 430 dead and more than 1,600 injured. It was the first mass raid on London (and the last in daylight) but it heralded the first of fifty-seven consecutive nights of bombing.

Mothers and their children were evacuated all over again, not in a mass but in 'a daily stream piloted to areas of relative safety according to the circumstances' in an evocative post-war description. Families were sundered, pets abandoned while whole streets were burning.

There was a renewed rush by pet lovers to the lethal chamber – but this time with no official encouragement to do so. A 23-year-old Mass-Observation diarist, Miss M. Rose of Grays, Essex, noted:

> Tonight I have been registering animals for NARPAC. Very few people have ever heard of it, although one or two have trained their dogs to go to the shelters with them. At least 30% of animal owners say they will have their pets destroyed if the air raids should get any worse.

At the time Miss Rose was writing, late July, bombing of towns had not even started and in fact NARPAC had been strongly urging do-not-destroy-your-pets through the spring. Individual expressions of anguish are much harder to find than during the mass slaughter of September 1939.

For humbler Londoners, this time it seems to have been a case of outright abandonment of their animals. Christopher Pulling reported a despairing phone call from Colonel Stordy in the early hours of 18 September: 'Dogs left behind in the East End have strained all their [NARPAC] resources, all the accommodation is full and they have had to slaughter wholesale. They are applying to the Home Office for assistance, particularly in the disposal of the carcasses.'

Mr Pulling could report that the Dogs' Home, Battersea had responded with the minimum of fuss and bother and had found extra vans to round up East End strays. Harrison, Barber & Co., however, were being difficult in the disposal of carcasses. And so the PDSA would report, 'again we had to open our grounds for the receipt of their bodies, this time receiving a further quarter of a million animals. Our Technical Officers will never forget the tragedy of those days.' But this time it was more commercially organized.

In the enormity of civilian animal suffering to come, why reprise the oft-told story of a wartime pet who later found international fame and was awarded a medal? Because in a narrative dominated by dogs this story concerns a cat, and one who was there from the very beginning. It goes like this:

In 1936 a skinny, stray female London cat took refuge in the church of St Augustine's and St Faith's on the edge of St Paul's churchyard (the surviving tower remains incorporated in the Cathedral Choir school).

Three times the church's verger took the cat out – and three times she came back. Father Henry Ross, church rector, decided she should stay. He named her 'Faith'. Four years later she gave birth to a black and white kitten, a tom, which was named 'Panda'. After the Sunday announcement in church, the congregation sang 'All Things Bright and Beautiful'. On 6 September Faith abandoned her basket on the top floor of the rectory and found a crevice near the crypt from which she would not be moved.

On the night of 9 September, German bombers targeted London in a mass attack. The East End was hit hard. Fires were burning around St Paul's and both sides of Ludgate Hill were ablaze. Altogether, over 400 people were killed and 1,400 injured. St Augustine's Rectory received a direct hit. The rector was in a shelter but next morning he hurried back. As a popular post-war account put it:

> His home was a mass of burning ruins. His first question to the firemen was to ask news of his cat and kitten, only to be told that both must be dead.

> He disregarded their orders to keep out of danger, and climbed over to a parapet from which he could see into the ruins where the remains of the recess in which 'Faith' had made her home should have been. He called the cat's name, and to his unutterable relief heard the kitten's faint mew ... Sitting in it, serene and unafraid, he saw Faith with her kitten between her paws.

They were both completely unhurt. A picture of Faith was hung in the tower chapel with an inscription: 'Our dear little church cat of St. Augustine and St. Faith. The bravest cat in the world. God be praised and thanked for His goodness and mercy to our dear little pet.'

That was the funny thing about cats. They were amazing survivors. And like Faith, mother cats were observed many times performing heroic acts of maternal devotion by moving kittens to perceived places of safety, one at a time. Did cats have some higher power? Our Dumb Friends' League noted in their 1940 report: 'Cats have some sixth sense which humans do not possess, that warns them of danger, and many casualties have been avoided by the fact that they have been able to escape before the house has actually been struck.' That, or they find some crevice in which to stick it out.

Bombs demolished the Glendale Road premises of the Hampstead Society for the Protection of Animals, for example, 'but six cats which had been in the Shelter all survived unscathed, apparently having sensed trouble and sought shelter, although nothing was left standing'.

Anglo-Indian Mrs Garbo Garnham (née Mander) was aged five in the Blitz, living with her mother Princess Sudhira Devi of Cooch Behar, in a pet-filled house in West Hampstead. She remembered her mother, 'boiling sheeps' heads for hours in the garden.' Her older sister, Gita, drove an ambulance and would come back each night with bombed-out cats and dogs. We got homes for most of them,' Mrs Garnham remembered, 'but one little biscuit-coloured puppy had a fit and just died on the spot.'

Seeking respite in Birmingham, her mother, the Princess, swept into the Grand Hotel and demanded her entourage of pets be accommodated. They were.

Air attacks on British cities would continue for seven months. It was not to be NARPAC's finest hour. There was a gathering cash crisis, bills were going unpaid and letters went out to regional vets imploring them to lean on their local authorities for financial assistance. As the bombing intensified, it fell to the charities to do what they could on their own initiative with their own resources – and still without real official recognition. Animal rescuers were the last to be allowed on to bombsites.

The Duchess of Hamilton's freelance rescuers simply jumped in their ambulance and drove through the streets looking for animals in distress. One of them, the extraordinary Miss Rita Cannon, would record soon afterwards:

> An old man stopped me with tears running down his face, saying his dog, after a severe raid, had been running wild for days. He had seen him on several occasions, but the poor, frantic creature would not come to him. Some children and two men offered to help me.

They found the dog after half an hour in the debris of a house. 'I decided to come back later, alone,' Miss Cannon wrote. 'After trailing the dog over debris and through cellars in the end I was lucky, with a lasso.'

The Duchess was meanwhile scooping up refugees for her Wiltshire sanctuary. She rattled through the cases:

> Dog belonging to Mr. B. Telephoned us. His house had been bombed, his wife, child and six-year-old black Retriever, 'Jack', had been buried under the debris. Mr. B. had dug the dog out, bruised but alive, 45 minutes after the Rescue Squad had given him up for dead. His wife and child were now living with him at new address.
>
> They were going to the country on 5 November. Jack was tied up in the yard, which was not satisfactory. All the other societies had said, 'have the dog destroyed,' but Mr. B. said, 'If the dog was meant to die he would have died under the debris.' He was placed by us in a good home in the country.

And so it went on. There was the case of the Dalmatian bitch, 'Trixie', whose owner was 'very poor, and unable to keep her'. There were two male Borzois whose owner's house had been bombed, wife and family evacuated. 'Blackie', a Cocker Spaniel, was the subject of complaint by the horrid air raid warden for general nervousness under fire. And there were plenty more cats. As the Duchess wrote:

> There were some very special cats, such as 'Timoshenko' [in January 1940, Semyon Timoshenko took charge of the Soviet armies fighting Finland in the Soviet-Finnish War] so named on account of his great courage and daring. He was one of twenty-four sad and miserable cats who arrived one cold, snowy night at Ferne – soon after a very bad raid in the City.

> A few days after he arrived he escaped from the cattery and joined on to a pack of about thirty-eight dogs large and small – who were taking their afternoon walk (Timo knew no fear). He is now an institution and sleeps curled up in the arms of a young spaniel and in the mornings they wash each other's faces with great solemnity.

The RPSCA too sought to get pets out of the firing line. Its 1940 report told how 'many balloon units' had asked for cats as mascots, and also as a way of keeping down rats. 'The Society saved quite a number in this way from being put down owing to the death of their owners, or because of their home had been destroyed.' An RAF squadron was presented with 'two Siamese cats' by the wife of Inspector Quigley.

The Whitechapel Shelter of Our Dumb Friends' League was in Venice Street, a little way north of the East India Docks and in the front line. Its end-of-1940 report recorded: '19,864 cats, 2,255 dogs and a considerable number of fowls, pigeons, rabbits, birds and guinea pigs have been collected or brought to the shelter,' most of them since the air attacks on London had begun in September. The tales were harrowing:

> One dog was brought in by the police with a piece of shrapnel through an eye, another was found in the Mile End Road, badly cut by a bomb splinter, a cat was found in some ruins, having lost one hind leg and the lower half of his jaw, a dog was found wandering in the East End, but he had his name and address on his collar showing that he came from north west London, having been bombed out of his house, and travelled in his fear, all that distance.

> Another dog was rescued after being buried for two days, but his experience was too much for him, he developed hysteria and was humanely put to sleep.

Then there was the story of a cat who was evacuated from Bethnal Green to Dagenham, and 'after an absence of three weeks found his way back to his old home, but as the owner of the property who resided on it kept birds and refused to have a cat, so he had to be put to sleep.'

Every day the League's Whitechapel staff found a host of 'deserted animals, their owners have gone away with no thought at all for the creatures living with them'. But there were some Londoners who clearly felt the very opposite. The story was told of two elderly sisters living in a bombed street in a working-class district of south London, along with two Pomeranians, 'Bimbo' and her daughter 'Flossie'. 'Neither of them is afraid of the guns,' explained Miss Fanny Dyer, 90. 'We never go to a shelter. I believe we are the only people on the street who stay put. When the raids are on, our dogs snuggle up close and we don't mind for if a bomb came down, we should probably all go together.'

A Mass-Observation interviewer covered Shoreditch in late September. 'Animal pets are frequently regarded as important members of the family group,' she noted and told this story:

> One man for instance preferred to stay at home with his cat and rabbit rather than to accompany the family to a shelter, although his house had been cracked across by blast, and said: 'I know the missus and kids are safe so I don't worry about them, but if I went to the Shelter too I'd be thinking all the time about how the animals were getting on.'

But fleeing owners presented certain opportunities for abandoned pets. A Birmingham housewife recalled for author Norman Longmate 'the shock of emerging from the shelter to discover that the cat had eaten the weekend joint'. A Bristol family recounted how they were starting on a frugal supper of cheese and biscuits when the sirens sounded: 'We all dived under the dining table as we heard the first bomb fall, and on emerging found our dog had polished off all our suppers. He was licking his lips, and looking not the least ashamed.'

Cats showed a universal will to cling to the wreckage. 'Many piteous tales of cats seen roaming the ruins of their homes have reached us,' reported Our Dumb Friends' League. They appealed to the public via local newspapers to 'take them home and wait for us to collect them'. But would-be rescuers just let the cats go again when someone failed to turn up pretty sharpish. 'It is obviously impossible to collect cats in the dark,' bewailed the League. The volunteers had little or no petrol. This well-intentioned initiative seemed doomed.

One of the worst conundrums was the delayed action or unexploded bomb (unforeseen in ARP planning), which meant whole streets might be roped off for days with pets locked in houses or roaming piteously in search of food. There are many accounts of amateur rescuers defying orders not to help them.

The long-established Hammersmith ODFL Shelter at Gordon Cottage, Argyll Place, stayed open during raids as a place for people to leave their pets whenever they themselves sought cover – with the express wish 'to find new homes should anything happen to them'. 'Sixteen dogs belonging to these unfortunate people [killed by bombing] have been so placed,' confirmed the League's end-of-year report. Meanwhile:

> A monkey had been left in a house while his owner went to a public shelter. The man was killed and the monkey would not let anyone go near him. After some time the staff of the League contrived to get him under control.

The 'no-pets in public shelters' rule proved highly contentious. Neither were London pets allowed in the LCC-run shelters for bombed-out families, nor into the Tube system, unofficially colonized from the start of the bombing, although smaller animals were smuggled in bags or beneath winter clothing. ARP Wardens had their orders. There was much heated language (and barking) at shelter entrances. I know I could not abandon my own animals and wartime pet owners were clearly no different.

In Manchester that October, Mrs Carrie Constance Hewitt was summoned for taking her dog to a shelter during an air raid and allegedly, 'becoming abusive when asked to leave it outside'. When the offended ARP Warden gave his evidence, the magistrate, a certain Mr Pugh, had clearly already decided the prosecution was trivially vexatious. He asked whether it was forbidden to take 'panthers or camels' into the shelter. When the warden replied that it was not, the case was dismissed.

It had been noticed from the beginning that dog walkers in parks were reluctant to leave their pets on the sounding of an air raid alert. The Canine Defence League sponsored a shelter for posh dogs in Kensington Gardens. On 26 September the first brick was ceremoniously laid by Sir Robert Gower, MP, accompanied by Sir Charles Souter, Chief Air Raid Warden for Kensington. The League hinted darkly at why it had taken so long, 'because of the crass and blind attitude of people who work in secret in order to indulge their own petty prejudices'.

'This air-raid dog shelter is not a great inspiration or a marvellous achievement in itself, but is rather a symbol of the sympathy which Britain extends to its dogs in a time of stress,' said its proud sponsors. And it was. It would prove popular too.

On the same fashionable side of London, a First Aid Veterinary Post had been established at Animal Defence House amid the elegance of St James's Place with an air raid shelter for pets in its basement. There were signs posted in Green Park advertising its presence to Mayfair pet owners. It was under fire from the start.

There was the case of a little cat who had been 'a conspicuous and attractive inhabitant of a well-known restaurant which suffered disaster,' according to the Duchess of Hamilton. The Society's veterinary surgeon, Mr F. C. Holliday-Potts, searched for the creature, which he found severely wounded – 'He was able to give a peaceful death, in his car, to this pathetic creature.'

The Mayfair post was supposed to cover Westminster but its veterinary nurse, Miss Rita Cannon, clearly went to wherever she might be needed. Her adventures were remarkable. They were recorded in a dramatic pets-at-war Blitz diary published the following year:

> Everywhere the poor folk crowded round me, asking me to take dogs and cats. Many of the cats were so wild that they were difficult to catch, but the children were all helping me and, in the end, I filled the ambulance with animals.
>
> I have been up to [blank] Road with buckets of food and drink and went from street to street feeding the cats. As soon as I appeared they rushed at me twenty and thirty a time. They all but spoke. I have since paid two more visits to this district each time bringing away

the cats. All I could do for them was to give them merciful release from their suffering.

Each organization did what they could. The Camden Town branch of Dumb Friends' reported the case of a dog found after being buried for six weeks in rubble: 'He was a pitiful wreck, and they did not have many hopes of saving him, but with devoted care he pulled through.'

In Sheffield after the attacks of 12–13 December, the People's Dispensary rescue squads had to 'clamber over ruins to collect cats where the owners were dead or injured. Thirty-one animals were dealt with, alas mostly to be destroyed, through owners forgetting to take their pets when they sought shelter,' according to *PDSA News*.

Rescuers found an old lady in a tiny house, the owner of a little dog with a cut foot. 'He is all I have now, please help me do something for him,' she implored. 'Tears were close at hand.'

In London, the Superintendent of the Our Dumb Friends' League's Chelsea branch in Bywater Street reported a cat buried in his basket in Pont Street at the back of Harrods. 'Such is the cleanliness of these animals that this cat had washed himself constantly and when rescued was without a speck of dirt,' she noted. She was called out again by the Royal Engineers to a half-demolished school where a 'very wild bitch' separated from her puppies was in the utmost distress – 'She had also been wounded and it took several days of coaxing before she could be caught.'

The Bywater Street superintendent[17] next reported a call for help when a blast smashed the doors of cages in which

17 The freehold of 20 Bywater Street had been left to the League by Mrs Lily Susan Brockwell of Chelsea, who had died on 4 December 1938, on condition it was used as an animal shelter. That is how many pet shelters were established.

chickens and rabbits were kept at the convent of Adoration Réparatrice in Beaufort Street. Seventy-two chickens and twelve rabbits were rounded up. The Mother Superior was advised that the accommodation was insufficient and the animals were evacuated. A few days later the empty coops received a direct hit.

The Chelsea nuns were among the many new urban smallholders who had answered the Government's call. Domestic livestock keeping had turned out to be a mixed blessing for pets. Goats provided milk, while rabbit meat could feed dogs and cats (but watch out for intestinal parasites). But that was really meant to feed humans. The trouble was that it was all too easy for the smaller 'economic animals' to become family pets themselves. Could you eat the one you loved?

The official Domestic Food Producers' Council Rabbits Committee had noted the 'sentimental aspects' at its first meeting on 27 March 1940, 'that rabbits in small numbers in different households come to be regarded as pets and people are loath to kill and eat them'. And furthermore, people in some districts 'preferred the taste of wild over domestic rabbits'.

By now, after over a year of war, rabbits and chickens were a part of city life. Labour Leader Clement Attlee, Deputy Prime Minister, kept a goat called 'Mary' at his Stanmore, Middlesex, home. The genteel sounds of the wartime suburbs were supplemented by grunting, meh-ing, clucking and crowing. Not everyone was happy, though. The neighbours of a Middlesex woman objected when she installed two dozen chickens in a garden run – half of them cockerels given to loud crowing at dawn. But they were soon 'wholly won over being given three eggs a day in rotation and the promise of a chicken for Christmas, and were soon willingly contributing scraps to feed the birds'.

Egg-laying chickens might be spared the chop for a while but rabbits had only one way to go. A single tame rabbit could yield 2 lb of meat.[18] The skin could get you money off the rag-and-bone man.

Official guidelines for amateur poultry farmers were to have new pullets (young females) every year, which meant the older generation would then be heading for the oven. Were they pets or dinner? Mr F. G. Imm of Stafford recorded the dilemma.

> At the end of their first laying season our birds were in a condition fit for the oven. [But] how could we eat Brownie, Spotty, Hoppity or Blackie after they had given us so many eggs to supplement our rations? I did not have the courage to wring their necks so I asked the milkman to do it for me.

Some families were tougher-minded from the start. In one memoir: 'Chickens were killed for food as soon as they stopped laying eggs. We would all go and watch Dad wring a chicken's neck, and then we would sit in a circle and pull out the feathers.'

Avril Appleton, a wartime evacuee from Hull, hated their backyard poultry. They attracted rats, 'and once there was this dead rat in the nest and [my parents] made me get it out'. And not all rabbits were lovable. She remembered:

18 Recipe for Rabbit Pie

Prepare a rabbit and cut into joints, Make a mixture of a tablespoonful of flour, a teaspoon of curry powder, salt, pepper and a pinch of mace. Roll the pieces of rabbit in this mixture then place them in a pie dish. Pour in enough stock to keep the pie moist and add a few neatly cut root vegetables. Cover with suet pasty and cook in moderate oven for about 2 hours.

The Field, 1942

> Then my mother got these rabbits that were called Flemish Giants, for the meat and so she could make clothes from the fur. Awful things, horrible great big things, you couldn't make a pet of them. But when they did get killed and mother sent it away, it cost her no end getting this fur cleaned and all it made was a small pair of mitts for me, which I promptly lost.

Norman Longmate's epic collection of Home Front memoirs recorded a Liverpool factory worker whose cousin kept rabbits – 'One day one was killed and prepared by the butcher. And when dinner was served – 'one by one the plates were pushed away untouched'.

A Hull woman who had 'never skinned a rabbit or plucked a hen in her life' quickly learned how to do so with the aid of *The Smallholder* magazine. 'I never ate any of our own rabbits,' she told Longmate, 'but my husband enjoyed them.'

'Phoebe', 'a large, fat, white rabbit', nearly escaped the pot when a Dagenham schoolboy discovered neither his mother nor sister could face consuming her. At last he was sent with the rabbit to an aunt in Southend, 'where Phoebe was speedily converted into stew'.

And in an Internet memoir, a Southend woman recalled a family wartime story: 'My mum (as a very young girl) was told that their pet rabbit had run away with the cat next door. She found out some years later that they'd actually eaten it.'

Eating pets was hard enough for grown-ups. Telling the children what was really for dinner was impossible. Or you could always blame Hitler.

Chapter 14
The Comfort of Pets

The compassion shown by humans for animals under fire was tangible. It fitted the heroic narrative of the Blitz generally but the animal charities were keener still to promote pets themselves as heroes. The vet H. E. Bywater, the likely writer of the anonymous West Ham chronicle of the great pet-killing panic of September 1939, reflected on the mood of a year later after intense bombing had begun:

> Readers may conclude that animals were an unjustifiable liability during the Blitz on our town but that would be unfair. Many a tale could be told of animals giving aid to their owners, of dogs leading them to shelters before the siren sounded or as bombs had fallen, or leading rescue squads to trapped persons or standing guard over the bodies of their departed friends.

He reflected too on how animals kept up morale. 'Many a weary person has found solace and courage from contact with a pet,' he wrote. That is how it was for him:

> The writer has an old ginger cat. Night after night in the early days of the war, the writer and his wife, tired out

> after a long day, crawled back to an empty house for the children had been evacuated. An empty house? No! Ginger was always there to welcome them and keep them sane amid the injustice of man's mad warfare.

The companionship pets gave, especially to women whose husbands were called up, the reassurance they offered in the blackout, were all cited whenever anti-pet sentiment erupted, that dogs especially would bite or go berserk in air raids. A housewife's Blitz testimony given in *The Dogs Bulletin* provides a good example:

> I am suffering through raids and am having severe turns of collapse and am often alone till midnight, as my husband is a bus-driver. When coming out of these attacks of mine, I find the little female pup, aged 10 months, banging her little nose on my knee. She will not leave me until I am all right. Dog-haters should learn what friends our dogs are. Well, are they a danger in raids? To me they are a comfort.

But the dog haters were as loud as ever. To draw their sting, in October the Canine Defence League announced a splendid scheme to collect dog-hair combings, 'which will be spun into yarn by voluntary corps of spinners and thereafter knitted into service comforts'. The spinning would be done 'by Highland crofters under an agreement made with Miss Dorothy Wilkinson, co-principal of the London School of Weaving'. How could dogs be thought unpatriotic now?

Were pets comforting humans or was it the other way round? Mrs C. Tavener of Crouch End in north London had taken in the kitten of 'Mrs P.', her bombed-out next door neighbour. But every time the little cat heard the children's calisthenics programme, *Music and Movement*, on the wireless, the 'swish' sound that the presenter made to tell bemused children they were in some scary storm-battered forest, meant the 'kitten seemed terrified'.

Swish, swish! 'Every time there is a swish sound, her eyes went round and her ears pricked up,' wrote Mrs Tavener. 'Is it her terror at the fall of the bomb and the crash of the house?'

The kitten had decided to use the Mass-Observation diarist, 'as her refuge' and 'won't leave my lap and sleeps on my bed at night,' she wrote. It seems an ideal arrangement to me.

Some pets even sought to aid and comfort other pets. RPSCA rescuers reported cases of 'cats going back into a burning house to find their kittens and giving their own lives in the endeavour'. There was the story of an inspector who 'went to a house soon after a raid in search of a stray cat'. He searched everywhere but could find no sign and was just about to leave when 'he suddenly saw a furry form dive beneath the rubble'. He discovered four ten-day-old kittens with the stray nursing them as her own.

In another sentimentalized (nothing wrong with that) account, an RSPCA Inspector recalled the rescue of 'a very frightened and bedraggled tabby from a ruined building'.

> But puss would not leave the debris. She even tried to re-enter the hole through which he had just pulled her, so the Inspector made further excavations and found a little black dog, badly wounded, crouching there in the bomb hole. It was gently lifted out and tended, to the evident satisfaction of the cat.

Like the story of Faith, the mother love of pets was especially moving. The RSPCA reported a heroine cat who carried her kittens one by one in her mouth from an upstairs bedroom in a bomb-damaged house, by way of the telephone wires to a place of safety in the garden shed.

Dogs too. In October, the story of 'Nelly', buried for five days in the rubble of her home, was in the papers: 'Underneath her, covered by her body were her five puppies, she whined now and then to guide the rescuers.' Mother and pups were alive and well – 'Starving herself she had fed her puppies throughout.'

The People's Dispensary was perhaps the keenest to promote hero pets. *PDSA News* told the story of 'Beauty', a Wire Haired Fox Terrier belonging to a dispensary caravan driver, Mr Bill Barrett. On their emergency call-outs, Beauty 'stands with her paws on the edge of the window next to the driver's seat, barking with approval. With her keen nose down she hunts everywhere and as soon as she detects a buried dog or cat, her barking soon brings the rescuers to the spot.' *Dog World Annual* gave her star treatment:

> Her sharp senses will often detect the presence of a dog or cat in the midst of the debris and she stands on the spot 'pointing' and giving off a volume of barks until somebody starts digging. Beauty has more than twenty rescues to her credit, and some of them even include cats!

Beauty would be the People's Dispensary's poster dog until final victory. It was reported that 'admirers presented her with little leather boots to protect her paws which became raw and bleeding in her efforts to reach the victims of air raids'.

And then there was 'Rip' whose exploits were not so publicized at the time (but who would later become very famous). After his Poplar home was demolished, the Jack Russelley-something attached himself to a bespectacled

ARP Warden called E. King. 'No one knew how old he was,' when he wandered into the Southill Street warden's post, homeless and hungry. 'He can't have been young,' recalled Mr King later. 'After blackout Rip always accompanied me on my nightly rounds of the street shelters. He was always sure of a titbit from one or other of the occupants.' Together this team would become exemplary workers in the rubble of east London before anyone had officially thought of using dog power to find victims in the ruins. Little Rip was Rescue Dog Number One.

There were plenty of tales of miraculous human survival down to pets. Take 'Fluff' for example, described by the *Daily Mirror* as a 'modest little dog with the heart and courage of a lion'. Fluff was buried with her owners in the rubble of their house – 'By continuous scratching, she managed to make a hole big enough to scramble through, which also acted as an airway for the people still trapped. Then she stood outside the hole and barked until help arrived and her family was safe.'

An Airedale bitch belonging to a dairy company horse-keeper made Lassie-like insistences that a family sheltering under a kitchen table should go down into the cellar before the house collapsed on top of them seconds later. 'Teeney Weeney', a brave tabby, fluffed herself up to enormous size and frightened off a looter.

The home of 'Peggy', a 'terrier of sorts' was shattered by a bomb and her female owner trapped under debris – 'The baby, Peggy's special charge, was in her pram, she worked furiously with her paws till she had made a hole through which the child could breathe. It is no small tribute to her skill and patience that all three – mother, child and dog – were saved alive.'

How could such pet heroes and heroines as these be regarded as in any way dispensable?

Chapter 15
How is Your Pet Reacting?

From the earliest days of planning for war, the expectation had been all along that pets would go barking mad at the first wail of a siren.[19] 'How is your dog reacting?' asked *The Dogs Bulletin* at the end of 1940. 'On the one hand it has been stated that dogs and cats take little notice of bombs and gunfire, while the other picture presented is one of distracted hysteria.'[20]

'Cats are comparatively indifferent or even contemptuous to the efforts of the German air force,' said the *PDSA News*. 'They return to their homes the next day and sit among the ruins purring quite contentedly,' while, 'birds remain cheerfully singing in their cages in shattered rooms without

19 The businessman and philanthropist Alexander Duckham wrote to the Prime Minister in 1941 with a suggestion for changing the Air Raid Warning note from 'the demoralising wail of the amorous gargantuan tom cat' to 'cock-a-doodle-doo' signalling 'defiance and triumph'. Mercifully for pets it was not adopted.

20 Author's note: From my own experience of my two beloved rescue cats' reactions to firework night in London, I personally believe this is attributable to individual personality. One cat, 'Fergus' (a beautiful cream Maine Coon), although normally the more timid of the two, shows little or no fear of the explosive bangs and sparks, while his co-habitee, the bold black 'Luis', is reduced to cowering under the gas cooker.

walls or roof.' One London office cat curled up in the in-tray when the sirens went, a favourite snoozing spot normally denied him.

'My bees are frightened out of their wits every time the anti aircraft guns fire,' a correspondent for *Bee Craft* noted in October, while a month later an apiarist reported, 'a bomb made a fabulous crater a few feet behind our twenty or so hives. A few minutes later my wife and I were surprised to find the bees working away quite unconcerned.'

As for dogs, there were stories on both sides. When a bomb fell 200 yards away, a 'large' Retriever 'crept under a sofa and without a bark or yelp expired,' recorded *ARP News*. 'A pack of startled greyhounds ran wild through a town for days until eventually they returned to their kennels,' the same edition reported. The majority of dogs of a provincial city 'abandoned it entirely' according to another report, 'to wander the countryside for days.' Stories of such mass panics however seemed apocryphal.

George Orwell's diary for 17 September recorded reactions in his Marylebone mansion block; its inhabitants huddled with their pets in the basement:

> All the women, except the maid, [were] screaming in unison, clasping each other and hiding their faces, every time a bomb went past. The dog subdued and obviously frightened, knowing something to be wrong.

His own dog, 'Marx', a Poodle, was 'also like this during raids, i.e. subdued and uneasy,' he wrote. 'Some dogs, however, go wild and savage during a raid and have had to be shot. They allege here, and Eileen [his wife] says the same thing about Greenwich [her family home], that all the dogs in the park now bolt for home when they hear the siren.'

All sorts of patent anti-hysteria dog pills were being pushed in the press. Canine columnist 'Philokuon' advised: 'Some people have bromide ready to give the dogs when the alarm is sounded. In such a case it should be given at once to let them have time to steady down before the racket begins.

'The dose varies from 3 grains to 20 grains in a tablespoonful of water, according to the size of the dog. Those that are kept in outside kennels should be brought indoors, for, if they are left to themselves, they are bound to be scared.'

But were such measures really necessary? *ARP News* could report after months of bombing that if 'a really well-trained dog is kept out of view of flashes' then all those nerve tonics and ear wads being promoted were not needed.

Instead dogs seemed positively serene under fire. Some indeed were heroes. It was the owners who panicked first. *The Animal World* reported the view of a Midland vet who thought fears of hysteria had been greatly exaggerated. 'A timid and excited owner may transfer his terror to his pet,' he said – 'after visiting a house following a frantic telephone call, it was not the animal that was in most need of a sedative.'

'Animals are very sensible,' said the North London Home for Lost Dogs, 'and dogs have apparently got used to the noise. There have been lost dogs, but some are lost due to [the] owner's carelessness.'

The Dogs Bulletin loved stories of dogs keeping calm. When an aircraft was brought down near Croydon, it reported, 'the engine became detached and dropped through the roof of a kitchen belonging to a nearby house. It went straight through the roof and on to the boiler. This in turn exploded and blew a hole through the wall. A dog that was in the kitchen at the time calmly walked out through the hole, unhurt.'

The naturalist Eric Hardy wrote: 'My own Collie sheepdog has been in the shelter with us without showing the least disturbances or barking once. He has become ARP conscious, preceding the family into the shelter at a warning and then with raiders past signal getting up and returning to his kennel.' However, Hardy was very tough on less resolute dogs:

> Badly trained dogs however have proved a nuisance, barking loudly and often rushing wildly about the place without entering the shelter. Muffed ears just add to the excitement and they struggle to get rid of the wadding.
>
> If the feeding of dogs becomes a national grievance, these should be the first to be destroyed. The barking of one dog sets off all the badly trained dogs in the district.

Plenty more dogs were apparently developing an 'ARP sense' with the ability to distinguish between the alert and the all-clear signals.

As did cats. 'At the alert she comes indoors to take shelter,' said a loving owner in response to a newspaper request for information on pets' reactions, 'but when the raiders-passed signal is given, she jumps up and scratches to be let out.' Some pesky dogs seemed to *like* the noise of guns and bombs, while others were 'miserably frightened and crawl under the furniture'.

'Some parrots definitely dislike the noise, and scream loudly and hysterically,' it was noted. Songbirds in the wild generally treated aircraft as if they were 'hawks', the birds 'scattering downwards and crouching to avoid detection'. Neither robins nor swallows paid any attention to aircraft.

The reactions of animals to war would continue to fascinate observers through the winter and spring. It was remarkable how little it affected them, even at the Zoo.

Captive wild creatures did not possess powers of premonition. 'There is certainly no evidence that any of the animals in the Zoo can anticipate the arrival of aircraft, distinguish between ours and theirs, react in any important way to the sirens, or foresee the impact of bombs,' wrote a 'special correspondent' for *The Times* soon after the bombing began. 'The animals are, however, not entirely neutral to disturbances. After a few minutes of anti-aircraft fire the other night the cranes got rather excited and began their rattling cries; but they soon settled down.

'The gibbons have a cry that resembles the siren in some ways although nobody who lives near the Zoo fails to distinguish them.'

On 27 September the first bombs fell on Regent's Park. The Zoological Garden's Occurrences Book noted:

> 38 incendiary bombs. Extensive damage to main restaurant by fire and water. Gardens closed until further notice due to unexploded bomb. Zebra and Wild Asses' house both damaged beyond repair. One Grevy's Zebra and wild ass mare and foal escaped but were captured next morning. Bomb in road at back of rodent house. Animals uninjured.

Twenty-four rhesus monkeys were noted as having been sold by the Zoo to Dr Zuckerman for '£12 the lot'.[21]

21 That September the anti-vivisectionist press drew attention to an article in *The Lancet* by Dr S. Zuckerman of the Department of Human Anatomy, University of Oxford, on the effects of blast on rabbits, monkeys and cats. Exposed to 70 lb of High Explosive at eighteen feet, only two out of ten cats were killed. The Professor was

In a December talk on the BBC director Julian Huxley would tell the story of the night when the Zoo was blitzed. 'By a miracle none of the occupants was killed or even hurt, apart from a few bruises and scratches,' he recalled. 'The female wild ass got out of her stable into the paddock, while her foal escaped at the back and wisely took refuge in the stoke hole under the hippo house, where she was found next morning.'

> A zebra [called 'Johnson'] was liberated into the gardens and made his way through the tunnel and out through the gate, which the fire brigade was using. He headed for Camden Town. Keepers set out in pursuit and by a combination of coaxing and driving he was without much difficulty got safely back and into an empty store shed.

A senior Ministry of Home Security official wrote in October to regional officers about 'unease in the country owing to the presence of exotic animals (lions, bears etc.) which may escape as a result of air attack'. There were also known to be reptiles, big cats, etc., kept as fashionable pets. He was canvassing opinion. Should special powers be sought that they be evacuated or slaughtered? Were they indeed a danger? It seemed they were not. Like the nation's more conventional pets, their main concern seemed to be where the next meal was coming from.

working to a Ministry of Home Security Research and Experimental Branch, contract with bomb effect trails conducted at a disused brickfield at Stenbay, north of Bedford. This presumably is where the Zoo's monkeys were heading.

Part Three

PETS SEE IT THROUGH

'I Help My Pals'

Motto of the Tail-Waggers Club

Chapter 16
The Perils of NARPAC

The beasts of England could take it. Muntjacs and mongrels, Pekes and penguins ... There were no panic-stricken animals rampaging through the country. So what was NARPAC for? After a month of bombing, by October 1940 the crisis had come. Committee members were at war with each other while a delayed-action bomb in Gordon Square had caused operations to be suspended. Creditors were demanding their money. The number of Animal Guards had fallen from 47,000 to 16,000 due, it was said, to 'boredom and local quarrels'. Talk of a great army of three million guards had been fantasy.

With the committee in uproar, the chairman, the former Ministry of Agriculture Principal Assistant Secretary, Mr H. E. Dale, was 'sorry it had come to this,' so he told the Minister of Home Security on 26 October. He complained, justifiably, that the Government had brought the welfare charities together and then given them no status. They were now also a registered war charity, to the horror of the RSPCA, alarmed by the threat to its jumble sales and stately flow of legacies.

'The government cannot wash its hands of all responsibility for dogs and other pets,' wrote Mr Dale.

'The government controls everything and is responsible for everything [true enough by late 1940]. Domestic animals may not be a material interest of the nation but they are a strong sentimental interest.'

According to him, the animal welfare societies were hugely jealous of their own status while the veterinary profession was 'suspicious of everyone'. The Vets' Association meanwhile thought that NARPAC's lack of discipline was the problem – that, and the inability of 'well-meaning individuals who love animals to appraise the problems sanely'.

Some vets had thought it better to join the Home Guard or regular ARP, the better to serve animals, so ineffective had NARPAC proved. Henry Steele-Bodger, president of the National Veterinary Medical Association, accused the constituent welfare societies of 'wallowing in publicity for their own self glorification'.

'Their leaders are without conscience and their promises are worthless!' he declared. 'The benefit of animals in wartime was not the prime concern of certain members of the committee.'

The Minister was further told that the RSPCA's lack of co-operation was based on jealousy of the registration data and against the PDSA and especially Our Dumb Friends' League on the grounds that they were 'extremists'. The Society's supporters would be most unhappy if the 'normal flow of legacies and subscriptions [were] interrupted or turned towards other animal welfare societies competing with it for public and private favour'.

For a while NARPAC looked as if it would simply collapse under its own contradictions. 'I shall not be sorry to see an end to the whole unsavoury business and I doubt whether pets would be any worse off,' so Mr Eric Snelling, the official now responsible at the Ministry of Home

Security, told Christopher Pulling. But instead they commissioned an outside report (it is unsigned but it is by someone 'hitherto unfamiliar' with the committee). With the Blitz at its height, its commissioning was a reflection of political concern, for good or ill, for the fate of Britain's pets.

Its conclusions on the war between animal lovers were damning. 'NARPAC has no legal existence or signed articles of association,' the report reminded the Minister. 'The crisis had been brought by dissension and financial difficulties due to a lack of confidence in the future.' It had been a mistake to ask the animal charities to pool ideas and money and show, 'more kindness to each other than kindness to animals can apparently evoke'. Furthermore:

> Animal welfare charities are especially competitive for publicity and funds. The animal loving public is volatile, blood sports are an issue between them.

The charities remained true to their Edwardian instincts – craving quasi-royal patronage in the bosom of respectability. That is where the legacies came from. Now the Government was the only patron that mattered, already smothering every aspect of national life in socialistic officialdom. Milk for children, if not yet for cats.

The anonymous invigilator had interviewed the principles separately – 'to ascertain the strengths of the jealousies'. It was a fascinating insight into animal-lovers' motivations.

RSPCA committee member Mr Arthur Moss 'undoubtedly wished NARPAC to fail'. 'The RSPCA is not really interested in 'pets' and 'it regards the PDSA and ODFL as upstarts,' said the Ministry's investigator.

Sir Charles Hardinge (chairman) and Mr Edward Healy Tutt (secretary) of the Dogs' Home, Battersea were 'not interested in any of it. Their principle concern is the disposal of carcasses of destroyed dogs.'

The municipal vet, Mr Bywater, was 'quite unfit for the task of financial controller'. Meanwhile Mr Johns of the NCDL thought that Bridges Webb was an 'outright gambler who had manipulated the registration scheme so the PDSA could buy it up'.

It was Edward Bridges Webb who emerged as 'the motivator and by far the most interesting of the representatives,' said the report. 'All the solid and potential assets, as far as "pets" are concerned, are tightly controlled by this man. He is master of the situation and we have no option but to recognise the fact.'

The Ministry called him in personally on 1 November for an interview. He was frank about the schisms, especially between vets and the People's Dispensary – which the profession had always regarded as 'unqualified quacks' (the RSPCA further had brought pre-war prosecutions against their staff for 'cruelty' when performing simple surgery). Besides, vets were only interested in large animals in the country. In towns, their surgeries were in the better-off districts. Of the 120 animal welfare clinics in London, 'very few of them were in poorer areas,' he said.

The pet registration scheme had failed to cover its costs, he admitted. It had raised £23,000 (£1m+ in today's money) but it had cost the same to implement. There had not been much concentration on cats 'because of the cost of collars'. The scheme could be expanded after the war and it paved the way for the licensing of cats (a long-term political ambition of the 'radical' animal societies).

Mr Bridges Webb made some extraordinary admissions. He had designed the NARPAC badge and had had it registered at Stationer's Hall. Legally it was his property. He himself 'had taken over the staff at Harrison, Barber & Co.

and the Merton Bone Company', and was 'now disposing of animal carcasses at a profit by reducing them to their by-product.' At least the money was going to a good cause.[22]

Harrison, Barber & Co.'s West Ham rendering factory had itself been blitzed and 'could only dispose of the bodies of dogs three times a week'. The Dogs' Home, Battersea had resorted to burying hundreds of animals that it had lethalled in 'a large hole' in its own garden. 'It is essential that continued pressure be brought to bear to obtain a dump or series of dumps within a reasonable distance of London,' the secretary, Edward Healey-Tutt, told Scotland Yard on 1 November.

Five days later he confessed to having 'climbed out of further difficulty' by disposing of the carcasses at a secret site at Longfields, near Dartford in Kent. The local authority did not know about it, and there would be 'considerable trouble' if they did.

Food and petrol were short. He was getting horsemeat from Harrison, Barber & Co. but was now eating into reserves of biscuits. He was meanwhile 'finding homes outside London for any dogs worth keeping but a large proportion are puppy bitches, which nobody wants, or diseased mongrels'.

By mid November NARPAC's cash situation was critical. Trying to salvage blitzed farm animals was difficult and costly. With grim humour, a country vet in Hampshire, Mr J. F. Tutt, demanded some sort of official incorporation into the ARP so as to 'prevent an undertaker being presented with a dead cow and a butcher with a specimen of homo sapiens for salvage'. He despaired at the futile attempts to run the organization on some kind of 'quasi-military basis'.

22 War was good for business. *The Times* recorded, 'Harrison, Barber & Co. (animal slaughterers) – A profit of £8,607 is returned for 1940 (against £3,471 for 1939).'

By now more than 4,500 'Parish Animal Stewards' had been appointed and 3,500 food animal carcasses salvaged. But pet welfare groups were subsidizing it all. Farm animals were not their concern. A total of £90,000 (£3.95m) was needed immediately. Creditors were clamouring. When the issue came to a head at a conference of regional veterinary officers there was a motion that the Government should be asked to bail them all out (the ODFL and PDSA voted against). But still the Treasury said no.

The travails of Britain's 'Ministry of Pets' reached neutral America. *The New York Times* reported the refusal of dogs and cats to react to bombing in the way His Majesty's Government thought they would. 'Instead of going mad, they adapted themselves quickly to new conditions,' wrote *The NYT*'s London correspondent. 'The noise of sirens bothered them at first, but a few days after the start of the Blitz they began to understand what the noise meant and led the family procession to shelters. Similarly horses took the noise as a matter of course.

'Herr Hitler's introduction of the time bomb, however, has created a big problem with thousands of people forced to evacuate their homes, unable to take their pets with them.' The same problem was true of those evacuated from 'invasion corner' on the south coast, 'the problem now is to find a temporary refuge for hundreds of thousands of pets.'

'Special animal shelters have been built in parks and private canteens have been set up to feed stray dogs,' he wrote. 'While the British Government does not like the idea of feeding dogs food that has to be convoyed into this island, the British people in their usual humanitarian fashion are seeing to it that as few pets as possible are destroyed.'

Clearly the American journalist had been briefed by insiders on what he called 'The National Pet Register'. To function it depended on telephone calls to the capital when it was often impossible to get the call through. 'Consequently hundreds of thousands of pets had to be destroyed and officials are now trying to meet this problem by decentralizing their organization and letting each city take care of its own register,' he wrote. 'The problem of financing this new machinery came up before the officials today.' This was a carefully spun version of the truth.

Nevertheless, New Yorkers were treated to a vision of a nation of humane pet lovers to whose aid neutral America must surely come:

> The trouble to which the British people and officials have gone for their pets is simply staggering. Hundreds of voluntary workers, including veterinary surgeons all over the country, are giving their services free of charge and risking their own lives every night to take care of these animals.

The top vet Henry Steele-Bodger found all this bending of the truth most unpalatable. He burst into a rage on 4 December about a BBC broadcast, 'in which certain of the animal welfare societies have been wallowing in for their own benefit and self glorification, something the veterinary medical profession cannot do'.

He was finding the animal welfare advocates increasingly wearisome. In communications with the Ministry, Mr Steele-Bodger was by now questioning whether their love of animals had gone beyond reason.

*

The second Christmas of the war was coming and the intermittently fed goose was not getting especially fat. Backyard poultry breeders eyed up 'Henrietta' and 'Blackie' for the chop. For Britain's pets there were mixed reasons to celebrate the festive season under fire. Perhaps the charities would bury their differences in the spirit of goodwill. Our Dumb Friends' League led the way by hosting a Christmas party for bombed-out dogs at its headquarters near Buckingham Palace.

Mrs Maria Dickin published a moving tribute in the Christmas edition of *PDSA News*: 'That plaintive cry, the faint moan,' she wrote, 'how it rings in the ears and tears at the heart of the PDSA staff who, often at great personal danger, toil hour after hour to rescue the animals trapped under the debris.'

Mrs W. Slater of Birmingham was one of them. 'I have rescued hundreds of cats,' she told the *Daily Mirror* in early December, 'even while a house is burning, a cat will remain there. I throw the lasso cord over its neck and drag the cat out.' Her fiercely independent Harborne Home and Dispensary for Animals was the designated NARPAC detention post. 'Many of them are so badly injured they have to go into the lethal box immediately,' she said, 'and last week three lorry loads of dead cats were taken away from the city' [the heaviest attacks on the city were in late November]. Mrs Slater and her family had voluntarily become vegetarians, it was reported, 'so that the many cats she has rescued can have meat at her home'.

There was more upbeat news as well. During the Christmas Blitz on Manchester (22–24 December) it was reported: 'At midnight, a goat was found by the police at the Odeon Cinema in Oxford Street. This bewildered animal had luckily been registered with the National A.R.P. Animals Committee, and a telephone call soon resulted in restoration to its Longsight owner'.

Mancunian Animal Guards seemed terribly keen. After the festive-season raids they were reported to be touring the city on foot and by bicycle organizing relief for the 'wild, starving cats' of the city. Scraps of meat and fish from caterers and abattoirs had been collected and mixed with dried cat food stockpiled by the committee.

'Thirty volunteer women reported the cats' whereabouts and distributed baskets of food to approved animals,' it was reported. 'They carry rope lassos to catch stray dogs and cat waifs are brought to the lethal chamber.'

Mr J. T. Beilby of the Committee's North-west Region said, 'These cats are wild and elusive and we have to leave the food out for them to take. But I am convinced it is the cats who get it and not the rats.'

Since the Christmas 1940 raids, 500 cats had been destroyed by their owners and the same number more saved from 'hungry vagrancy' (by being destroyed by PDSA mobile squads), so it was reported. It was perhaps better for Manchester cats to stay clinging to the ruins if they were to survive.

In a round-up of evacuee tales, *The Dog World* told the story of a woman in a provincial town who broke down in tears because no one would take her in with dog and her baby – until 'the kindly owner of a local boarding kennel offered free shelter for the Westie [West Highland White Terrier] thus simplifying her search for lodgings'.

Cats too charmed their way into places of warmth and safety. *The Cat* told the story of the church cat of St Magnus the Martyr in the City of London: '"Puss" placed her four young kittens in the manger with the Bambino. A little lad visiting the church during Christmastide saw the kittens and with much delight shouted: "Baby kittens with mother Jesus!"'

And in Chester Cathedral during the festive season another correspondent saw 'a huge black Tom curled up very cosily upon the straw within the manger. He was not the official church cat, but may have been invited in by the resident puss – one never knows.'

Chapter 17
Non-essential Animals

In spite of all its troubles, by the start of 1941 the protecting veil of NARPAC had been spread over the whole country. Britain's cities continued to take a pounding as the pet lovers' militia now stood sentinel, supposedly, over *all* the nation's animals. Its committee members had been released from pending bankruptcy meanwhile by the intervention of the Ministry of Home Security. All sins had been forgiven; the Animal Guards were back. Whether this would actually help animals under fire had yet to be seen.

That early spring, a Ministry of Information photo session for the US press featured Mrs Olive Day of Drayton Gardens, Kensington in: 'A Day in the Life of a Wartime Housewife', where she was seen knitting a balaclava and preparing for her naval officer husband to come home on leave. Her agreeable black cat, 'Little One', accompanies her throughout, prominently wearing a collar with its NARPAC badge. The collar 'was to ensure, should he stray in Blitz or blackout, he will be returned safely to his owner'. How thoughtful.

There was a triumphant rally of Animal Guards in Manchester on 13 January to mark the first anniversary of the North-western region. Captain T. C. Colthurst, general secretary of the Registration Branch, was introduced as 'the original organiser of animal guards' and he told how just sixty guards had been registered in October 1939. Now there were three million nationwide, he boasted. It was an astonishing figure and entirely untrue.

'Throughout the country, 5,000 animals a week are being returned to their owners,' Colthurst claimed. He told the story of 'Dusty', a cat from Leigh-on-Sea in Essex found by a guard in Bloomsbury with a NARPAC collar and how within a few hours his delighted owner had learnt by telegram that his pet was safe. How the Essex cat had got to central London was inexplicable, but the registration system worked admirably.

There were complaints from the floor that in some areas police and ARP were obstructing rescue work for animals, but the Chief Constable of the city, Mr John Maxwell, was known to be sympathetic to pets (he was indeed a PDSA member). The number of cats still being destroyed was deplorable, said Captain Colthurst. Only recently he had 'seen an advertisement saying that there was a big demand for cats'.

By now the urban Animal Guards were joined by their rural comrades, the Parish Animal Stewards. Was there really anything for them to do?

The *Yorkshire Evening News* told the stirring tale of how war 'came viciously' to the East Riding when a 1,000-acre farm found itself in the firing line and sticks of incendiary bombs fell on stables and cattle-sheds.

Farm foreman Tom Swift, 'un-tethered the horses one by one, and let them gallop to safety. With the help of other farm workers, he released 80 head of cattle, and then, twice thrown from his feet by terrific explosions, he ran to save the farmhouse,' recounted the story. 'Tom Swift did no more than many a townsman has done since the Blitz came to England, but this is written so that posterity might know of the equal heroism of country folk.'

An anonymous correspondent for the *Manchester Guardian* sent a graphic despatch from the Lincolnshire fens 'where farmers have found that homeward-bound German bombers like to jettison any remains of their cargo before they reach the sea'.

A bomb dropped at Pyewipe Pasture had 'killed two good plough horses, and the farmer's proper down about it'. The NARPAC vet had given gas and raid instructions to all the cowmen and stablemen of the neighbourhood. A farmer from Mintin's Ear was the Parish Animal Steward, ready with his team of local butchers 'to proceed to any farm where badly injured beasts need to be destroyed'.

Pets had not been overlooked either. The village chemist had been 'besieged for the powders recommended as a sedative for noise-frightened dogs and cats. The transformation of packing-cases into gas-proofed dog-boxes has become a favourite weekend job.'

The blacksmith's wife was a canary fancier. The couple's son had brought back three such birds from Belgium, 'through the inferno of Dunkirk, ransacking abandoned shops for bird-seed to feed them.' Now, acting on instructions from the vet, a blanket soaked in hypochlorite stood ready in the parlour to shield the cage of 'Blitz', 'Gort'[23] and 'Victory' from gas attack.

'In many kitchens you will find pinned to the wall one of the leaflets issued on war-time feeding for domestic pets, badly needed now that dog biscuit is a luxury and table scraps are consumed at table and no longer discarded,' wrote the anonymous correspondent.

23 Named for Lord Gort, the Commander in Chief of the routed BEF. Pets generally were being named after personalities and sometimes places in the news.

'The totemic NARPAC identity discs had been dispensed by the rector's daughter with instructions to ink them in with the animal's name and owner's address. The decoration is highly popular and is being worn by the postman's rooster and Mrs. Yeats's two nanny-goats as well as by the whole canine population and three babies.

'We shall take good care of our animals, if only to prove the veracity of the Frenchman who said, "that in England it is lovely to be a dog".'

Such bucolic harmony was not universal, however. When it was announced that rationing of livestock feed would start on 1 February 1941, the gloom had deepened. 'Owners of urban horses including mules, asses, jennets, and donkeys,' it was announced, 'must now use ration cards to get their nosebags replenished.'

Domestic animals were not to be included in the livestock scheme. There would be no ration books for pets. Sir Robert Gower, MP, the RSPCA Chairman, complained in Parliament. He was told with some logic that 'A rationing scheme for dogs and other domestic animals would be impracticable owing to the great variety of breeds and the extent to which these requirements are usually met by purchased foods and by scraps respectively.'

They must continue with what they could get, but strictly no human food and rely instead on the many diets being promoted by the charities and by NARPAC too. The observer of Fenland pets had further written: 'I fancy these leaflets, with their excellent charts on the quantities of protein and carbohydrate for different breeds, will do something towards reducing the waist-line of those unlucky animals whose owners think of a dog as an ever-open mouth for titbits.' That might not be the general opinion of dogs.

On 1 January 1941 had come a new humiliation, an order

'that all meat which is unsuitable for human consumption must be dyed green'. 'The special dye is stated to have no harmful effect upon the animals to which it is fed and the colour cannot be boiled out of the meat,' it was reported. It was all deeply unappetizing.

Out in the countryside, the huntsmen of England faced deepening anguish. Some were realistic. A former Master of Fox Hounds declared that the food being eaten by hunters should indeed be given to cattle – 'Five couple of hounds should be the maximum held by any hunt; these should be kept by farmers and fed on refuse. The rest ought to be put down.' But a diehard replied that the time to destroy hounds was when 'all pet dogs had been made into glue'. That day might be closer than he thought.

Miss J. A. Boutcher, master of the Courtenay Tracey Otterhounds of Andover, Hants, admitted to the Ministry that her pack (now reduced from seventeen to ten couple) consumed the same as Foxhounds. Could she please have more food for them?

Major Cecil Pelham, the Hon. Sec. of the Masters of Fox Hounds Assocation, wrote to the Minister of Agriculture to say that: 'Seven thousand eight hundred hounds have been put down since the start of war. Masters and hunts have done their best to play the game.' And, in the context of what had happened to urban dogs, they had.

Thus it was that much-diminished hunts carried on through a second wartime winter without giving up entirely. There were fewer cars in the countryside to spoil the scent but also fewer open fields over which to ride. Every scrap of land was being put under the plough. Fox coverts were being buried under aerodromes. Tenant farmers, emboldened to defy their hunting landlords, simply went out and shot foxes.

But it was the Government machine in Whitehall that

would decide the fate of hunters and hunted – as indeed of everything else. Farmers and landowners had lost their last few freedoms when Britain became a siege-socialist state after Dunkirk. The Ministry's County War Agricultural Executive Committees ('County War Ags') had dictatorial powers to decide land use, issue 'plough-up' orders, direct what livestock would get what to eat, to direct gangs of labour (including Land Girls) and to eradicate what (and how) it decided were 'pests', all backed by emergency legislation.

On 6 February memos flew to and from the Ministry of Food exiled in Wales ready to brief the Civil Defence Committee meeting due to be held the next day in London on the question of animal feeding. The Principle Assistant Secretary teleprinted: 'We suggest that hunting should be stopped, it is impossible to justify.' And furthermore what about racehorses? With U-boats[24] slashing at the Atlantic lifeline, some big gesture was needed to let the public know just how serious things were. Officials asked how could the embattled island propaganda play in neutral America when horse racing merrily continued?

It was all about perception. Not much later it would be noted that, 'The general public, who for very good reasons cannot be given the true figures [on dwindling food stocks], are not going to believe the seriousness of the situation if race meetings continued.'

It was recognized that pain must be felt by everyone. How could huntsmen have their sport and parrot lovers and urban pigeon fanciers be denied theirs? 'I emphasise the political importance when all imports of bird seed have

24 The commander of the U-boats, Admiral Karl Dönitz, was said to be an ardent dog lover who, 'on his return home, his first greeting was always for the family dog, a little Spitz named "Purzel"'.

been stopped and there will soon be no seed for caged birds in private ownership,' one official told the Minister.

Class resentment was raised in Parliament, when racing pigeons, 'the sole source of the working man's recreation' cannot be fed, while the rich man could keep his racehorses happy. As well as pigeons, the working man liked race meetings. Fox hunts however were a different matter.

Munitions worker Mrs Elsie Barnes wrote to the new Food Minister, Lord Woolton, on 21 January from Meriden, Coventry: 'Why is the hunt allowed to trample down a field of turnips and beans? Food for the plebeian people and don't farm animals need food too?

'Communism is the answer,' she insisted. 'It is certainly steadily growing in our factory, thanks to the hunting set and their kind.'

If hunting pink was a red rag in Coventry, the continuation of horse and greyhound racing outraged others even more, especially poultry keepers. Dogs generally annoyed a correspondent of *Farmers Weekly* concerned with sheep worrying. 'There are far too many dogs hanging round villages,' he wrote in January. They had to hang around somewhere.

But back in the city, Bonzo and Oo-Oo still had no official ration. All the Ministry of Food had come up with was a statement: 'Dogs and cats must subsist on the limited supplies available eked out by inedible offal, horsemeat and the like.' Ugh!

But then the Food Minister confused things greatly by saying blithely at a press conference on 18 February 1941: 'I think I am right, that if a kindly disposed person, instead of eating their meat ration gives it to a favourite dog, that is not illegal.' His civil servants fumed. Angry memos flew. The Waste of Food Order of the previous summer had not been tested by prosecutions, it was noted. But it could be. What about 'food not fit for human consumption that could be given to livestock but which instead goes to seagulls, city pigeons etc?' one asked. That should be regulated for. There must be more rigour all round.

Dogs were the problem. Not seagulls or ducks, nor racehorses even. The canine population was estimated at three million licensed and half a million more outlaws. They consumed 5 oz per day of biscuits and 3 oz of protein. How to get rid of them? Raising the dog licence perhaps, as had been suggested before, or reducing the grace period for strays on police hands. That was not nearly strong enough.

The Canine Defence League was in poor shape to do much defending. That February a Mass-Observation interviewer found their Victoria Station House headquarters down to a staff of two men and a secretary, doing what they could. The rest had left to go into the services. The clinics were busy but 'it's mostly cats ...'

'An increasing number of the public regard dogs as unnecessary,' said the canine defenders, 'because of the food question. That's what's uppermost in the mind of the dog owners and we're in constant touch with Government departments.'

Running away was no escape, especially for a London dog. The police were very keen on clearing the streets and pretty soon the van from Battersea would arrive at the station to collect the overnight strays (north London dogs went to the ODFL home at Willesden). Destruction at Battersea had peaked in 1940, with 17,347 dogs destroyed under police contract. The 1941 figure would be 11,446.

It was all too much for Edward Healey Tutt, the Home's secretary, who was found 'living in the paraffin shed' on the site, eating out of tins and was invalided out. The work continued. On 28 May his deputy appealed for deferment of call-up for military service of Herbert Alexander Collet, the home's 'expert electrocutionist'. 'We should find it extremely difficult to replace him ... and Mr. Tutt is recuperating in Norfolk,' she told Scotland Yard.

Cats were also being measured up for the chop. 'There appear to be no statistics on the number of cats in the country. It may be assumed however that it is greater than the number of dogs, 7–8 million,' so said Mr T. C. Williams of the Ministry of Food in a report on 'non-essential animals' made on 21 March.

He had calculated that if consumption per cat was 3 oz protein material per day, it meant an annual consumption of 215,000 tons. Each cat might additionally drink 1 oz of milk a day, making 18 million gallons of milk a year. 'The figure is probably on the low side,' he added. But would the destruction done by rodents eating foodstuffs in the absence of 'a cat equilibrium' as he put it, 'really be more costly than the existing cost of maintaining the cats?' He conceded that the experience of cats on protective work in warehouses, factories, etc. showed that, 'they are worthy of their keep'. But lazy lap-cats had better watch out.

Meanwhile the Cats Protection League had chosen its time well in launching the 'Tailwavers Appeal' to aid 'homeless and evacuee cats' – with half the funds raised supposedly going to pay for a 'Cats of Britain' presentation Spitfire for the RAF. They had better hurry up. The Spitfire Mk V, dubbed 'The Dogfighter', sponsored by the Kennel Club and announced in *Our Dogs* magazine the year before, was about to go into squadron service.

Cat lovers sensed danger. 'It seems possible that an attempt may be made to introduce a rationing scheme for household animals,' noted *The Cat* in February. 'Knowledge of the prevailing ignorance about cats rouses a fear that this will be unjust to them.' The danger was that the 'cats-can-fend-for-themselves' superstition would gain ground – especially that farm cats expected to catch mice could live on a saucer of milk day (which was now illegal anyway). 'We appeal to all our readers to be prepared to avert a possible danger to cats,' implored the editor.

Mr Williams's Ministry of Food non-essential-animal survey could also report that there were ten to twelve circuses in the country, only three of which had thus far made applications for rations. That represented twenty horses per circus consuming 200 tons of food per annum in total. Lions etc. were presumably being fed on 'unrationed roughages and meat'.

Circus and animal variety acts toured wartime Britain uninterrupted. Vic Duncan's 'Royal Scotch Collies' who had performed pre-war on the lawn of Buckingham Palace continued their amusing routines. 'Jim Della's Dogs' performed their canine antics up and down the country. 'Watson's Fox Terriers (Everybody's Favourite)' were about to appear in *Mother Goose* at Bournemouth when their trainer, Johnny Watson, died aged 99 on Christmas Eve. Such acts had been a staple of BBC Television, to the anguish of the animal welfarists, before it shut down. Who would have thought dancing dogs could be popular?

Some animal welfare ultras had been on their case for decades, such as the 'Performing and Captive Animal Defence League' run by the extraordinary Captain Edmund McMichael, who was always in trouble for pulling down posters or generally protesting. In June 1940 the Captain had petitioned King George VI that both circus cruelty be addressed and he, personally, should convey a peace proposal to Adolf Hitler. 'As nations allow animals to be treated so must they expect retribution,' he suggested. His peace mission did not get off the ground.

The Home Office got involved when a Mrs Stella Lief of Raynes Park complained to her MP about a variety performance she had seen featuring four cats revolving on a 'Blackpool-style' wheel with a Pomeranian dog standing on its paws. The cats 'clearly indicated fear,' she said. The act, the 'Royals' Famous Cats and Dogs', was tracked down on tour in Norfolk. Its somewhat elderly brother and sister proprietor, Rose and Harold Crick, were found to be registered under the 1925 Performing Animals Act. Having seen the 'turn', a police officer reported the animals clean and fit and there was no evidence of cruelty. The performing cats and dogs danced on, untroubled by do-gooders or enemy action.

Our Dumb Friends' League grew animated when Chapman's Circus (featuring 'Jan Doksansky's famous menagerie of lions, polar bears and Himalayan black bears') got into financial difficulty in March 1940. The League's report recorded, the 'animals, dingoes, monkeys and penguins were bought up to stop them falling into bad hands. Some were so ferocious that it was the kindest thing to put them to sleep after giving them a good meal.'

By March 1941 the approaching animal-food crisis was all over the papers. Drastic cuts in pig and poultry rations were predicted. The main blame, for now, was being aimed not at dogs but 'useless' racehorses. It was getting visceral. 'Unless this war comes to an early conclusion, their conversion into sausages is likely to prove their only real contribution to this country,' an Oxfordshire poultry farmer told *The Times*. 'One racehorse consumes the same as 125 hens a day.'

The number of racehorses had been reduced to one fifth of pre-war, it was pointed out in return. 'Millions of people derive pleasure from racing and they would prefer a minimum number of race meetings to half an egg each year.' The hen vs. horse controversy would rage through the spring. No politician could yet find the will to ban horse racing outright and turn studs of thoroughbreds into pet food, although the public mood could change.

George Orwell (Sergeant Eric Blair in the St John's Wood Platoon of the Home Guard that covered Regent's Park) wrote:

> 'C' of my section of the Home Guard, a poulterer by trade but at present dealing in meat of all kinds, yesterday bought 20 zebras, which are being sold off by the Zoo. Only for dog meat, presumably, not human consumption. It seems rather a waste. There are said to be still 2,000 racehorses in England, each of which will be eating 10–15 lb. of grain a day. i.e. these brutes are devouring every day the equivalent of the bread ration of a division of troops.

The rumours were swirling. Racehorses, dogs, cats even, would all be eliminated. There must be a statement. 'Lord Woolton's promise – your dog is safe for the duration,' so *The Evening News* proclaimed on 25 March. 'Contrary to rumours, the Food Minister denied reports he was to reduce the dog population. "I have not even considered doing so," he said. "I cannot allow them any food but I have no intention of preventing them getting the food they now receive".'

The same edition reported that the local authorities were preventing the feeding of pigeons outside Windsor Castle.

The war between animal factions could be played out at an emotional level. It suited the Government that way – if something *really* convulsive did have to be ordered. The German dog destruction propaganda had cleared the way. Let the partisans of pedigree over mongrel, dog vs. pig, racehorse vs. chicken, fight it out. As long as everyone understood just how beastly one had to be to win this war. Meanwhile horse racing and hunting continued at a much-reduced level.

The crunch 'Food Consumption by Dogs' meeting came on 1 April. All the ways and means to reduce it proposed over the past year were reviewed.

'There should be no interference with domestic dogs,' the Minister decided. And although the Ministry would like other 'non productive animals', greyhounds and racehorses, put down, such a move was judged to be 'hardly practicable'.

'Interfering with the man in the street's dogs' would be politically equivalent to 'muzzling free speech, closing music halls or pubs,' it would be summarized a little later. The effect on morale of 'doing away with all pet dogs would have a worse effect than the loss of a military campaign,' was the legal adviser's personal view.

Dogs had been reprieved. It was politics. As the store cupboard emptied Lord Woolton must try to keep everyone happy, even pets.

The latest ration priorities for livestock were announced in the Commons the next day, 2 April. The primary concern for animal feedstuffs was to maintain the milk supply for humans. Working animals would get just enough to keep them efficient – 'this applies to working horses, pigeons in the National Pigeon Service and cage birds used for safety work'.

'The quantity of feeding-stuffs for sport or recreation

has either been drastically reduced or cut off altogether,' said Major Gwilym Lloyd George, the Junior Minister.

'Pleasure horses' would get nothing during the summer months and would have to be kept on a maintenance basis. 'Fox hounds, beagles and harriers are receiving, from April, rations for one-sixth of the pre-war numbers. The manufacture of dog biscuits has been reduced to one-third of the pre-war quantity. No feeding-stuffs are being made available for cage birds or for pigeons outside the National Pigeon Service,'[25] said the Junior Minister.

There would be further cuts in chickenfeed. 'Poultry-keepers are advised to cull their flocks rigorously and to send to market any unproductive birds,' it was reported on 3 April, 'it seems likely that there will soon be a large supply of boiling fowls on the market and no eggs to be bought.'

The 'complete denial of food for budgerigars and canaries' was raised in the Commons. 'It seems a small thing, but these pets are greatly prized by their owners, and many have to be put out of existence,' said the member for Linlithgow, parliamentary champion of cage-bird fanciers.

The Cat meanwhile sensed a 'far-reaching motive' behind it all. It was being whispered that 'facilities for the humane destruction of pets are to be officially provided'. The imminent *compulsory* destruction of 'unnecessary animals' was another rumour. 'It is difficult to tell where the official dividing line between the necessary and the unnecessary pet could be drawn,' said *The Cat*. A third rumour was that the 'kill-all-pets section of the community

25 The voluntary avian corps by which fanciers made their birds available for military use in return for an official ration of corn. Many thousands would be employed.

are feverishly seeking ways and means of forcing their views'. That was not just a rumour.

While cat lovers were 'prepared to go to considerable trouble to provide their little friend with its meals' *The Cat* complained at the same time about the exorbitant prices being charged for horsemeat and fish heads – and retailers who refused to sell even unrationed food, *if it was thought to be for a cat*. Perhaps it was all part of the same, sinister conspiracy against cats.

Chapter 18
Pets Under Fire

Who would champion pets now? Their political fate was not the concern of NARPAC, which in the spring of 1941 was to go through yet another internal spasm. The Ministries of Home Security and Food pulled out their representatives on the Committee without public announcement. The RSPCA and Canine Defence League formally withdrew altogether. A canine defender interviewed by Mass-Observation said: 'There was this scheme for Animal Guards, I know some of members joined, it never really came to anything.'

The researchers concluded that NARPAC 'has evidently been one of those committees at which representatives talked a lot, disagreed a bit, no one did anything in concord but quite a lot separately'. It was a shrewd judgement.

The health of the Committee's chief, Colonel Robert Stordy, was failing (he was to die within a year 'of an illness aggravated by the strains of trying to run the NARPAC') and he could not stop the infighting, had he ever been capable of doing so. The chairman of the Dogs' Home, Battersea (which had never co-operated properly), the hugely grand Sir Charles Hardinge, would blame NARPAC's troubles on the Colonel being so 'aloof from this wicked world, you might put him down as a mid-Victorian varsity don'. In fact he was a devoted vet who had spent a lifetime aiding animals around the world. Just when they needed practical help more than ever Animal Guards continued to meaninglessly register pets.

Meanwhile the trial of animals under fire had not relented. The industrial Midlands and latterly Britain's port cities were continuously attacked from the air through the winter and spring of 1940–41. Liverpool, Bristol, Portsmouth, Plymouth, Southampton, Clydebank, Belfast, Hull, Cardiff, Newcastle-upon-Tyne and Sunderland had all seen intense pet dramas. The case of Hettie Mary Symons, for example. It was reported:

> In a case heard at a south-west England police court [almost certainly Plymouth] yesterday, it was stated that the defendant defied a constable and risked a time bomb to feed her cat. She asked a War Reserve Constable's permission to enter a cordoned area to feed her pet. He refused, but took the food to the cat himself. Then he saw Symons in the area. 'I could not let the cat starve,' she told the court. She was fined £1.

From March to April, Plymouth was plastered by bombing. PDSA rescuers were almost overwhelmed. A swan was found saturated in oil from the naval base's ruptured tanks. The city's newspaper reported: 'The rescue squad spent a considerable time in tending the swan, but it did not improve so it was sent to the local headquarters of the unit.' Here it was 'treated for a week and given many baths, and also had its legs massaged. When it had regained its strength the swan was taken to quiet waters in Cornwall and released.'

And further, 'many cats destroyed had terrible burns on their paws due to running over hot girders and stones'.

Under the headline 'Puss Won't Quit Crater', a newspaper reported the tragic story of a cat whose family had all been killed – 'Housewives take him away, only to find he slips back to roam the pile of debris and dig for his dead master, crying as he searches.' It was heartbreaking:

> 'Puss' is black and white, and lives on scraps brought to him by children. 'He is a lovely cat,' Mrs. Mary Kitchin said yesterday. 'But he will not leave the crater where his friends were killed.
>
> 'We neighbours have often taken him home and made him comfortable, but he runs back as soon as our backs are turned.'

'Blackie' was another cat somewhere in the north-east, whose family had all been killed by bombing. Somehow the cat had survived in a cast-iron kitchen range for twenty-five days to be discovered by a demolition squad who heard faint mewing from the wreckage. Too weak to crawl out, 'she snarled at her rescuers but was eventually tempted out by a saucer of milk'. She had, it was reported, 'made a splendid recovery in the care of Mrs. Raper, whose husband was in charge of the squad who saved her'.

A cat buried for five days was 'found alive but died from shock very soon after rescue' *PDSA News* reported from the front line. 'Another cat from the same house was rescued, only to come back to the ruins wild with terror and would not let anyone approach it. When last heard of, it was taking food left out for it at night by a PDSA worker. Probably, by the time these words are in print, the cat will be safe in our care.' One can only hope this was true.

The port city of Hull was heavily attacked between 9 and 15 May. At Rank's flourmill, 'terrified horses were rescued from a blazing stable. They kicked and screamed in terror, but could not be moved until the rescuers placed bags over the heads of the horses.' Again the individual charities sent out their own rescue squads. NARPAC barely figured.

PDSA News reported dramatically on the rescue work in Hull, hampered by the hated time-delay bombs, of releasing animals locked in cupboards or cellars (by their fleeing owners) with no chance of getting food or water. One Technical Officer found, 'a crying cat in a blasted block of flats, it had been quiet for some hours. I rescued it, still alive, a black and white Persian. Its hair was burned to the skin, eyes and ears burned. I painlessly destroyed it.

'Four days later, I rescued from the rubble in a badly bombed, working-class district, a black Retriever lapping from a burst water cistern. His mistress was killed and his master terribly injured.'

In another street were, 'cats, dogs, chickens, canaries, budgies and ring doves, trapped by a 500-kilogram time bomb'. He found a parrot alive after five days, drawn by the noise coming from under the rubble: 'Polly's all right, Polly's all right!'

According to *Our Dogs* reporting on the fate of Hull's pets: 'It is worth mentioning that officials of the Dogs Home and the RSPCA, together with other animal lovers, dealt with seven hundred domestic pets, not hesitating to enter dangerous buildings to rescue them.'

Dog World Annual went somewhat over the top in reporting the People's Dispensary's gallantry in its survey of 'British Dogdom' in 1941:

> It is their proud slogan that they 'Go Where the Bombs Fall' and no call for help is ever left unheeded. Rescue

> workers have entered burning stables and they have mounted dizzy staircases, against the advice of the police, in order to reach some cat or caged bird, which had been left shut in a cupboard by some thoughtless owner.

So, who were these selfless volunteers? Who were these brave midnight feeders of ferals and listeners for cries beneath the debris? They drew their fortitude from an earlier tradition of concern, one with largely a female face.

Without the mad cat ladies (and some were really mad) of late Victorian Britain, there would have been a much-diminished 'animal welfare' culture to meet this new time of trial. The pioneering work of these much-mocked-at-the-time women had gone round the world.

The independence of such women was another reason why animal welfare was so fragmented. Like cats, they walked on their own. Right at the beginning, in the first flush of abandoned pets, a Mrs Thornton went out on to the streets of Leytonstone at night, 'looking into churchyards and derelict houses for the victims of callous owners'. She told her story in the ODFL report: 'I noticed a large, old, black gaunt cat. He was starved and neglected. He sat upon a roof where I could never possibly get at him. I began to throw food out to him, and then I found that somehow he came into the factory at night, and I fed him secretly. Of course, he found the food and would wait hours for me to feed him.'

Kindly Mrs Thornton was, after some time, able to catch him. The following year the League would record the story of Mrs Lilian Lane of Regent's Park, north London, 'where she worked single-handed from her own home'. She was approached by 'the distressed owner of a lost black and white male Persian cat' who had no address disc – 'He did however have a mark on his nose.' For weeks Mrs Lane searched in vain. Then, in her own words:

> One day, very recently, a sweet natured woman, who feeds all hungry cats in the district, told me about a black and white Persian she was feeding. I made several journeys to the feeding spot but did not see him. However, one Saturday I spotted him and his nose and there was the black mole mark. I went and fetched the owner who was overjoyed and they are all happily reunited.

Mrs Lane, an Animal Guard from the earliest days, would feature in a newspaper article as the 'Florence Nightingale of Animals', going round north-west London on her bicycle with its NARPAC badges emblazoned on baskets fore and aft, which has 'become a two-wheeled animal ambulance'. 'In the Blitz I had panniers on either side as well for I had a load,' she said. The article stated that 'In the past year she had had 131 stray cats put to sleep and found homes for forty cats, four dogs and one bird. Two dogs and fifty cats had been returned to their owners.' She regularly 'sat up till two in the morning working on identity discs'. Good and kind Mrs Lane!

And it was not just women. *The Cat* told a cheerful story in February about cats who 'refused to leave their bombed-out homes'. They had found, 'a friend in London bus driver, Mr. Arthur Heelas, who has, himself, been bombed from his home. Every day he tours the district with food for his furry friends, who emerge from the ruins at the first sound of his footsteps.'

And there was, 'animal lover Mr. C. J. Searle, a businessman of Petts Wood, Kent, who approached the

League about opening stations both at his home and at his place of business'. In the space of a year Mr Searle had, 'rescued and put down three stray cats, lethalled a cat whose owner could no longer keep him, returned one dog to its owner and arranged for a dog to be sent to a temporary evacuation home. In addition, he took into his own home one cat and one dog, which he had rescued from bombed premises.'

Rita Cannon of the Animal Defence Society recalled: 'Someone asked me to go and see a poor old woman who lived in a tiny lodging where she kept a number of cats. I found her indescribably bedraggled in a room swarming with cats and kittens, which she had found homeless and befriended. Once convinced of my friendly intentions, she agreed gratefully to my coming back to fetch the cats. Most of these poor people will give their last crust to their cat or dog.'

Whether they liked it or not posh cats and alley cats alike were recruited into the feral colonies now taking over bomb sites and military camps, surviving from pigswill bins and cook houses. Humans, for some reason, had chosen to put all their scraps and leftovers into convenient containers *at the end of every street*. Best get there before they were emptied.

After one raid on east London, an RSPCA Inspector reported, '1,400 pets, mostly cats, found all apparently ownerless'. The ownership of some was traced but, 'many of them were of the type that had no specific owners and were fed communally by residents in blocks of buildings'.

'In one East End street it took three months to round up a hundred cats during early morning and late evening visits,' noted the Customs House and Plaistow District shelter of the ODFL. As *The Veterinary Record* observed: 'Cats are the most difficult to retrieve, and so long as they are not suffering or starving, the policy is to let them remain and, thanks to a kindly public, many of them are regularly fed.' Fed with what? Giving human food to cats was illegal – it took a special kind of cat devotion.

In November 1940 the *Daily Mirror* made a big story out of a 65-year-old widow, Mrs Caroline Roberts, who 'has fed hundreds of homeless cats in a heavily bombed district of London and every evening makes a fire in her sitting room just to make them feel at home where they doze after she has given them a good dinner of cats' meat, bread and milk.' Ministry of Food enforcement officers must have been apoplectic.

There are plenty such examples in the archives of individuals who took pet collecting over the edge of obsession and ended up being dragged out themselves from some improvised shelter overrun with starving animals to face prosecution for cruelty or for wasting food.

Such pet-human dramas happened in different circumstances. In October 1941 there was a report from Barnsley, Yorkshire that Mr Carter, the huntsman of the Pershore Hunt, 'rather than see the pack cease to exist had not put them down'. The court was told that it 'had been impossible to get quality feeding stuffs and the hounds had lost weight'. One emaciated Foxhound's corpse had been kept in ice by the RSPCA as prosecution evidence. He was fined £10 for cruelty.

Wealthy zoophiles such as Nina, Duchess of Hamilton might be considered pet collectors on an industrial scale. In describing a fund-raising gymkhana at Ferne animal sanctuary, it was later reported, 'an aircraft hangar was arranged complete with central heating for the cats and staff engaged to care for them under the personal supervision of Her Grace'.

The cosy cat-filled hangar was at the estate's aerodrome,

laid out by her son, pilot and RAF officer Douglas Douglas-Hamilton, the Marquess of Clydesdale (who became 14th Duke of Hamilton when his father died on 16 March 1940), who would fly round the family properties, just as he would the pre-war European capitals, in his personal Tiger Moth. The dashing airman had been a keen appeaser who had made connections with prominent Nazis, especially those who shared a love of flying.

In late 1940 the Duke was embroiled in an MI5-Air Intelligence fishing operation to flush out potential Nazi sympathizers and general German intentions. These manoeuvrings would eventually lead to the flight of the deputy Führer, Rudolf Hess, from Bavaria to Scotland on 10 May 1941, bearing 'a secret and vital message' for the Duke of Hamilton, the Duchess's son, whom he had briefly encountered five years before at the Berlin Olympics.

Having bade farewell to his pet Wolfhound 'Hasso', Hess, an accomplished pilot, parachuted out of a Bf 110 long-range fighter over southern Scotland. His intended destination was the airstrip that the Old Etonian aristocrat had created in the grounds of Dungavel House. Although he had a map with Dungavel marked in red, Hess missed it and ran out of fuel to land south of Glasgow. The next day, Hamilton saw Hess in Maryhill Barracks propped up in bed with a twisted ankle.

If the deputy Führer had aimed for the Hamilton spread at Ferne instead, he would have found himself nearer to London to make his peace overture, but the aerodrome was inconveniently full of cats. MI5, the security service, kept an eye on the goings-on. When Hamilton's name first came into the frame, an officer noted on the file: 'I remember Clydesdale at school, he was up to normal but only just. It seems that his family were attracted by the

Nazi atmosphere.'[26] He further wrote: 'The Duke's mother is a well-known crank. Animals! etc!'

That was MI5's view of animal welfare campaigners. In an age of total war they seemed like long-frocked relics of another age. Nina, Duchess of Hamilton, might indeed have been a crank but no more than the many hundreds of women, young and old, rich and poor, who were drawn into the fight to help wartime pets. Like newly married Phyllis Brooks, aged 20, who worked with her ambulance-driving sister-in-law at Our Dumb Friends' League Hammersmith Shelter. She recalled later:

> We had lots of voluntary people in those days. There were many well-off women at home who did charity work. Wherever we went to rescue animals there was always somebody close by who would have a basket handy.
>
> A lot of people cared. When there were lulls in the bombing, people drifted back to London, then some would come and offer a home to a cat.
>
> Mostly we tried to talk common sense to people to help them nurse their animals back to health. We were the last of the rescue people to get on to a bombed site and we had to carry an identity pass issued by the National Air Raid Precaution Animals' Committee. Quite often there would be animals wild with terror and it would mean many visits to the site to set traps. There were people who would feed the strays, but we were kept very busy.

26 It was noted on the file that the youngest of the Duke's four sons, David Douglas-Hamilton, had married Prunella Stack, youthful head of the Women's League of Health and Beauty. The couple had visited Hamburg in late 1938 for the Nazi Kraft durch Freude (Strength through Joy) sponsored Congress of Physical Fitness.

'We were all described as cranky, we animal welfare workers,' said Mrs Brooks. You can bet they were thought cranky. They did it because with a war raging, they loved animals with a passion. How cranky was that?

Chapter 19
Wanted – Dog Heroes

Month after month, the continuing Blitz brought plenty of new observations of how animals reacted under fire. Cats were becoming positively blasé.

A Birmingham cat, 'Polly', was observed taking her three kittens one at a time to the cellar of a house during 'a terrific barrage'. Polly reappeared – 'followed by another cat, which she had apparently persuaded to join in the comfort and safety afforded by the refuge'.

When Chelsea Old Church in London was reduced to rubble on the night of 16–17 April 1941, 'a small cafe, with a nice cat, was destroyed,' it was reported. The owner of the cafe went back the next day and called for his pet among the ruins. He saw a heap of bricks moving to find, 'a small cat ghost, so encrusted was the little creature with plaster and dust'. Taken home and bathed, the cafe kitten became itself again – but 'after its morning meal, the kitten went back to the ruins, every day, returning to its new home at night'.

'My own two cats are getting quite used to bombs,' wrote the editor of *The Cat*, 'unless they fall very near, when they rush on to our shoulders and cling closely, quite stiff with fright. They love to go out on moonlit nights but make for home as soon as the siren sounds.'

That was a universal observation. Cats generally were reported as knowing the alert warning (wooing siren) meant danger from above – and they took to the cellar. There was a story about a London tabby sitting peacefully in a garden, who when the siren sounded 'sprang up, ran full tilt, and took cover in a public shelter'. The recorder of this incident had 'heard many similar stories of dogs, but none of cats'. She commented sagely:

> The dog is a friend, enters into our life and imitates some of our behaviour, even to the extent of looking both ways before crossing a busy road. But the cat is of another world, and his ways are not ours. It is an act of condescension on his part to avail himself of a human safety device.

And one cat was said to warn its deaf owner of raids. And on the first wail of the siren, Mrs Holroyd's delightful sounding 'Suli' and 'Meru' would go to where their leads were hanging (Siamese, you see), ready for the walk in the dark down the garden to the Anderson shelter. 'Once there, they would settle in their specially prepared gas-proof box and go soundly to sleep. At the first sound of the All Clear they are awake and ready to return to the house.' What practical creatures!

There were many such reports of cats emerging from their hiding places on the raider's passed sound (a constant tone). And there was more. Such as the tale of a London cat 'that knew the difference between the British "planes" and the German "planes"'. This miraculous feline 'could hear the Jerries' engines' – described as an 'uneven dragging, bumping' noise – 'sometimes before the siren sounds'.

The Cat was not quite convinced of this yarn while Professor Julian Huxley of London Zoo did not believe it either but he admitted that those animals that take notice of aircraft are often aware of them well before any human beings. 'The large majority of beasts and birds pay surprisingly little attention to air raids,' he said, 'but a minority react in individual and often unpredictable ways.' In the fascinating matter of extra-sensory and psychic cats, *The Cat* commented:

> It is, of course, difficult to assess the accuracy of observation of cat owners, unless one knows them. Sometimes the involuntary excitement by air raids might tend to make them attribute some of their own feelings to their cats.

But *The Cat* conceded that felines were indeed very 'telepathic', instantly picking up agitation in humans. If someone nursing a cat gave a sudden start, 'the cat usually flies off her lap. It may be, therefore, that a cat somehow senses the attention, perhaps a little anxious, given to passing aircraft by its human entourage, when they suspect it of being an enemy, and so appears to notice it more than a friendly 'plane.'

*

Enemy animals were under fire. How were they reacting? A few RAF-delivered bombs had fallen on Berlin Zoo, killing an eland (a type of antelope). 'It seemed incredible that a zoological garden with its innocent animals could be seriously considered as a target,' wrote director Lutz Heck of his November 1941 baptism of fire. Well, look who started it. At first the elephants had been disturbed by the anti-aircraft guns but now they went to sleep, he reported. The monkeys were agitated however but the beasts of prey soon got used to it. 'Lutting', the Terrier, seemed to enjoy it, 'soon realising that the sound of the sirens meant he would be going out and he could resume his ratting'.

From the Cats Protection League headquarters in war-torn Slough, the editor of *The Cat* suggested the telepathic powers of felines meant they picked up the nervousness of their owners as much as audible signals of approaching danger. And a member of the Canine Defence League wrote about her four 'most intelligent dogs'. These dogs could speak:

> When Nazi bombers are miles away, we cannot hear them. But the dogs will rouse from their sleep under the table, go outside and whimper. Then they bark a little and come back and tell me, 'Missus, they're coming.'[27]

27 Speaking dogs were a wartime phenomenon. *The Dog World* of October 1939 had news of a talking dog in Scotland, a Springer Fox Terrier cross in Troon, Ayrshire, who said (in a Scottish dialect) 'Ome' when asked where he wanted to go, and 'Mamma' whenever he was asked who was going to take him there. Telepathic, mathematical, philosophical and talking dogs had been a feature of German 'animal psychology' research for decades. A Dachshund named 'Kurwenal', who 'spoke' by barking a certain number of times for each letter of the alphabet, became internationally famous prior to his death in 1937. An intelligent Fox Terrier – 'Lumpi' – was visited in Weimar by the Duchess of Hamilton and Louisa Lind-af-Hageby in 1937. Louisa gave a lecture on the subject in London on her return, with a list of 62 'speaking' animals, most of them dogs, some horses, and 'Daisy of Mannheim', the educated cat, who was capable of doing simple sums and tapping out a word or two when she saw fit.

The Hundesprechschule (dog speech school) was founded in 1930 by Margarethe Schmidt in Leutenberg, Thuringia. Professor Max Müller, a Munich University zoologist, visited in 1942 and

The Veterinary Record had a more scientific approach. 'Throughout the country, pets have ceased to be alarmed at bombs falling or guns firing,' it reported. Their 'acute sense of hearing, and the rapidity with which they take cover had spared them' as pets found, 'with unerring instinct, sheltered places from which they emerge free from harm.' The *Record* also noted a universal phenomenon:

> Cats bombed out seem to pick their way out of trouble, and wander more or less leisurely away midst the noise, dust and smoke, eventually returning to sit on the ruins of their former homes.

Horses too 'have calmly carried on,' said the *Record*. They had become used to random loud noises through mingling with motor traffic on the street. Even fire was no cause for concern: 'In the East End of London, 54 horses were removed from a stable through an avenue of flame without injury. From the blazing stables of a brewery, 82 horses were evacuated without mishap.'

No cows had yet stampeded under Blitz conditions although milk yields might be down – 'Cattle appear to take quite an inquisitive interest in incendiary bombs, and some have been observed nosing round the bombs as they burned.'

Pigs had 'slept sonorously' through an air raid 'although the roof over them had been completely wrecked' (in fact they had been in an abattoir awaiting slaughter). Every single pig from a devastated piggery in Scotland was rescued alive.

reported dogs making speech-like responses plus the presence of a cat, and that Hitler himself had accepted Schmidt's offer for her dogs to entertain troops. In 1943 their food ration was cut off.

But sheep were nervous under bombing and ran 'hither and thither over the pasture'. Some showed an acute nervous disturbance – possibly 'shell shock' or 'vertigo', according to *The Veterinary Record* – 'One farmer reported that his sheep went round like dancing Dervishes.'

'The herding instincts of farm animals accounted for many casualties,' according to *ARP News*, in which the countryman author, Clifford W. Greatorex, told the story of a fox which ran into a house and sought sanctuary behind an easy chair in the sitting room in a Midlands village – 'where he stayed until daybreak'.

The author of this yarn was also informed by a poultry farmer that, 'at the sound of an aeroplane, his fowls, with one accord, ran for the shelter of the hen house'.

Pets keeping calm and carrying on was winning them a place in the tableau of national resistance – that and the devotion of their owners. A newspaper columnist in summer 1941 observed Londoners queuing for pet food, 'ordinary men and women paying tribute to their friends,' as he described them. 'Such actions prevent one from losing faith in humanity. An isolated case of supposed overfeeding of animals should not be allowed to detract from [such] unselfish efforts.'

That summer, Mass-Observation noticed 'increased pro-doggism' in a London survey. It was, they said, 'entirely emotional' in that the feeling was based on pets being 'part of the family'. This was a 'predominantly female response'. What the Mass-Observers found was revealing all round.

By now dogs were getting over their fear of air raids, so people said. Mongrels 'took it' better than thoroughbreds, the strongest effects of air raids being noted by the owners of Dalmatians and Pekingese. Upper- and middle-class people had Alsatians, Spaniels, Retrievers, Setters and

Terriers but very few mongrels. More affluent people worried more about their dogs than less well-off folk.

A woman living in a poorer district of the capital said: 'When I leave him, he does kick up an awful row. I can hear him all the way down the street. I go to the tube [shelter] and I can't do anything but leave him. He mopes for days afterwards.' She thought 'holiday homes for dogs', some sort of communal provision to get them out of the cities, would be ideal.

Another London woman said: 'I have been alone in the house since my husband went and the children were evacuated. That dog's been wonderful company – I mean wonderful.' She added: 'I've no patience with these people who coddle their dog, feed them on chicken and champagne, and take them to bed with them.'

Mass-Observation tabulated the most popular names for dogs among poorer owners. They were in order: 'Nip', 'Bill', 'Bonzo', 'Pluto', 'Spot', 'Jock' and 'Rover'. The survey found that the work of 'NARPAC is not really known or understood' – and concluded to increase the popularity of dogs (for which M-O seemed to be campaigning on a market research contract from the Bob Martin company) – 'there should be more done to show the ways in which dogs are helping the war effort. There have been singularly few *dog heroes* in this war.'

Chapter 20
Pets on the Offensive

There were few dog heroes, found Mass-Observation. That could change of course, and it would. Thus far courageous pets had got on with enduring the Blitz at home, along with everyone else.

The prospect of dogs actually doing something warlike had taken a knock early in 1941 when Army testers looking for four-legged guards concluded that a dog would bark at anyone, friend or foe, who was not his handler. 'It would be impossible to train a dog to differentiate between a British soldier and a German parachutist,' said the officer in charge of 'vulnerable points' policy. It was a case of one man, one dog, but individual handlers could not stand guard forever. With thousands of freezing soldiers doing guard duty against invaders, GHQ Home Forces were keen to keep trying. Trials of 'patrol' and 'messenger' dogs, which would have to move around a bit between several handlers, looked more promising. Where to get hold of suitable dogs? In May 1941 a stirring press and wireless appeal was made to the British public:

> The War Office invites dog owners to lend their dogs to the Army. The breeds most suitable are Airedales,

> Collies (rough or smooth), Hill Collies, Crossbreds, Lurchers, and Retrievers (Labrador or Golden), although intelligence and natural ability will be the deciding factors in selection.

'Rex', the columnist for *Tail-Wagger Magazine*, would one day write of how it was for him:

> Well, pals, here's how it's started...
>
> My mistress was listening to the wireless one day, something about big dogs being needed. She got up all of a sudden and said: 'Rex, you've got to go out and do your duty, you're going to be a "sojer" dog.'

One feels sure that Rex had his paw on the button when he wrote that.

That August, the Leeds branch of the PDSA offered seven dogs. 'Mick', an Alsatian, did not know any special words of command, according to his auxiliary fireman owner, but 'he'll answer to Oi!' he said. 'Few of these dogs would have been transferred from being pets to the military but for the difficulty of getting dog food these days,' it was observed. The owner of 'Toby', a Spaniel, was on 'long hours and war work, and that's the main reason little Toby goes into the Army'.

Dog lovers offering their family pets were assured that, 'those not passing the test will be immediately returned. Selected dogs will be retained for the duration of the war.' Owners were invited to 'write to The War Dog Training School, Willems Barracks, Aldershot' (the administrative address – the dogs were at Woolton House, Newbury). And it was not just large soldierly-looking dogs: Cocker Spaniels and Pekingese were also on parade. Within four days the War Office was 'inundated'. That was enough dogs, it was announced, for now.

So pets, or to be specific dogs, were going to go to war, even if they were only on 'loan' ('lend-leash', as one clever journalist described it). But rather than waiting for invasion, there was a cockpit of war where British forces were actually on the offensive. And already there were pets in abundance there.

Since September 1940 the British had been busy confounding the Italian Army's attempt to invade Egypt. It did not get very far. The PDSA hospital in Cairo, so it was reported, was 'full of dogs which had belonged to the Italians and had been stranded during the ebb and flow of the fighting. Some of them were very well bred, Belgian sheep-dogs, Italian Pointers and Setters, the majority however were pariahs.' There were also numerous cats. In the way of things, they were very soon adopted as pets.

This volunteer army of domestic animals would henceforth accompany the British Army and Air Force in their epic desert campaign. Once selected and trained for war, the first pets from home would be heading out to join them.

In February 1941 the Commander-in-Chief, General Archibald Wavell, was ordered to halt his counter-attack from Egypt into Libya and send troops to Greece. The RSPCA launched a bid to intervene on behalf of Greek animals. Like the doomed missions to Poland and Finland, it did not get very far.

On 12 February an otherwise obscure German commander called Erwin Rommel arrived in Libya to stiffen Italian resistance and soon went on the attack. Commonwealth forces were trapped in the besieged port of Tobruk. Expelled disastrously from Greece meanwhile, British forces had to take refuge on the island of Crete, to be evicted again by German air landings and rescued by sea on 31 May. Adopted pet animals were with them all the way, according to the PDSA history; they were 'smuggled out in boxes and kitbags'.

British troops fell back into Egypt and moved meanwhile into Iraq and Syria, lest the whole Middle East be lost. In Cairo, as the PDSA's own history put it, 'batches of dogs appeared and with them foxes found in the desert. It was a common sight in these dark days to see lorry loads of our soldiers coming in bearded and dusty with their new pets riding with them.'

The story was told of 'one little dog which had been captured from the Italians, sent across to Greece, evacuated to Crete, from Crete to Cairo, then to Syria and Palestine and back to Cairo – where it arrived on a 15 cwt lorry with eight big infantrymen and a corporal'. This 'rough-coated brown and white dog of unknown breed' had had five puppies. The platoon had movement orders; their commanding officer had forbidden the little family to accompany them. One member of the platoon promised to return but the outcome did not look promising: 'Instructions were left at the PDSA Hospital that should no one come and collect the dog and her family after one month, the puppies would be put to sleep and find the mother a new home. The platoon did not return,' it was reported starkly. Another, homelier account has the 'little dog finding a home with another soldier' while her 'babies were painlessly put to sleep'. A REME corporal in Cairo had a cat, which he kept in his stores. Twice it became ill and twice after treatment it recovered. Finally it died before the corporal could get it to hospital – 'The grief-stricken owner cremated the body of his beloved cat. For weeks he was convinced that he had done wrong.'

There were two dogs 'belonging to Tommies,' Hans Bloom, director of the PDSA Clinic in Cairo wrote, which had been brought back from Dunkirk and had seen Wavell's campaigns and the Syrian campaign – 'They had changed hands many times.'

Superintendent Bloom painted a lyrical picture of kindly soldiers – British, Australian, Rhodesians – selflessly rescuing pets, easing the burden of abused pack animals, while 'puppies for sale in the streets of Cairo by boys are bought out of sympathy for ten piastres each'. A Polish soldier had adopted a young desert fox. A wounded Greek soldier was adopted on the battlefield by a German police dog. He had woken from unconsciousness in a shell hole to find the dog licking his face. After that they were inseparable. With the ding-dong battle in the desert, there would be many more canine side-switchers.

The desert also had charms for homesick huntsmen. With not much to report from the English shires the sporting press printed lyrical letters from far-flung correspondents. An anonymous officer wrote from a camp in the Iraqi desert, inspired by the sound of 'dogs circling in the darkness, their voices keening in song as they hunt the desert hares and jack[al]s'. His mind was taken off in reverie, 'to summer days spent by the streams of Sussex with the Crowhurst Otterhounds'.

*

The presence of cats could also provide comfort. Three days after the Crete débâcle, Winston Churchill was at Chartwell, his house in Kent to which he made occasional wartime visits. For some years now it had been the home of 'Tango', described as a 'beautiful marmalade neutered male cat'. Sir John Colville, his principal private secretary, was there for luncheon on 3 June as the Mediterranean disaster was still unfolding. He recorded:

> I had lunch with the P.M. and the Yellow Cat, which sat in a chair on his right-hand side and attracted most of his attention. He was meditating deeply on the Middle East. While he brooded on these matters, he kept up a running conversation with the cat, cleaning its eyes with his napkin, offering it mutton and expressing regret that it could not have cream in war-time.

All this excitement in the Balkans and the Mediterranean meant that the air attacks on Britain were winding down. The *Luftwaffe* turned east to prepare for the titanic assault on the Soviet Union. It began on 22 June 1941.

*

The fantasy world of NARPAC had relocated meanwhile to Mr Edward Bridges Webb's own home in Golders Green in the north London suburbs, where the registration scheme was now headquartered. The bizarre task continued. Three days after German troops had crashed through the Soviet frontier defences, he boasted to the Ministry of Home Security that 'with no statutory funding, the registration branch has enrolled some 47,000 voluntary workers, the National Animal Guard, who have raised £35,000, registered three million animals, and are now a part of urban life'.

He proposed turning the debt-laden registration branch into a limited company. But following the withdrawal of the Government's representatives, 'serious doubts have been thrown on its authenticity and indeed its honesty is now in question,' said Mr Bridges Webb. The members of his council feared a 'grave scandal'. At least the veterinary profession had stayed loyal.

So should the Government disclaim all responsibility? 'Although we could still have lots of trouble [and] in spite of these people's difficulties, they have, in fact, looked after pets,' so the Minister was informed by his officials. From the animal lovers' point of view, it was noted 'an unfortunate impression' might be created by disbanding the organization.

Mr Bridges Webb had a new wheeze: metal collar badges costing a shilling each to 'cover the cost of free veterinary attention and insurance'. The People's Dispensary's dream of a welfare state for pets might yet come true.

Mass-Observation's attitude-to-dogs survey could not have been more timely. Researchers asked: 'Do you find your dog has been affected by the war? Does he mind you leaving him when you have to take shelter?'

'I think it's made him rather nervous, but he's forgotten all about the raids now,' said one woman in north London. 'I give him a couple of aspirins, but he's got a lot older and doesn't like games anymore. Funny, isn't it?'

Another owner's pet was not so calm. 'He kicks up an awful row – I can hear him barking all down the street,' she said. One woman bent the rules for her pet:

> Well, I have a little shelter I go to nearby so I'm allowed to take him with me. He's so small no one makes a fuss. He curls up under the eiderdown to get away from the noise, but he never makes any noise himself, he just shivers all the time.

There was a lady with a Pekingese called 'Tinker Bell', who was 'so well mannered, I take her everywhere with me, and she's so small nobody notices her. I take her into restaurants, she loves that.'

Not all stories were so happy. 'I used to have a dog, a Bull Terrier. He was so sweet I called him "Sugar",' said an interviewee.

> I adored him. He loved me so much. He didn't really like anyone but me, but the trouble was he couldn't bear to be alone and when the war started I became an [ARP] warden. Well, I couldn't have him with me at the post and he started to mope.
>
> I felt he was so miserable, so I made up my mind to have him put away. I nearly died myself. I took him for a little walk, he was so happy. But it was the only thing I could do. He would have gone mad in an air raid. Poor Sugar.

But a non-pet owner in Cricklewood felt that: 'When a country is in great danger it's ridiculous to clutter up the place with dogs. They should be looked after by the Government or destroyed. Food is so short for ordinary people it isn't right to give it to animals who are for the most part useless.'

The interviewer concluded there was 'no really strong feeling that dog-keeping is unpatriotic'. The majority of people interviewed however, dog owners or not, felt the Government's attitude was anti-dog. But there was a strong feeling that whether or not people kept dogs in wartime, it was none of the politicians' business.

The food issue was the big one. Profiteering was rife. An anonymous female M-O diarist recorded: 'Rush to shop to get fish for cats. I'm offered cod's head for 2s 4d a pound, which I decline on principle (although I need it). It would have been 4d before the war or given away for nothing.'

At a pet shop in Neasden, north London, researchers found an expectant queue. There was plenty of horseflesh but no dog biscuits. One woman said: 'There's that notice up saying the green colouring [dyed so by law since the start of 1941 as not for human consumption] is perfectly harmless though people are very sceptical. They just wouldn't believe it isn't poisonous.'

The Neasden housewives had no inkling of course of the existential decisions being made at the Meadowcroft Hotel in breezy Colwyn Bay on the future of their pets. That spring there had been a deeply horrid moment when the Ministry of Food spelled out that it was 'a serious offence' to offer food fit for human consumption to a sick dog. Any vet giving such advice and any pet lover doing so would be liable to heavy penalties. Keith Robinson, secretary of Our Dumb Friends' League, waded in, asking 'in reply to inquirers to our society, am I to say that His Majesty's Ministry see no alternative but to allow such dogs to die?' What a heartless Government, denying food to sick pets!

There was a big policy conference on 23 July 1941 to discuss the position of so-called 'non-essential animals'. It was very sensitive and highly secret. Sir Bryce Burt, veteran colonial agronomist, director of the Feeding Stuff Division, was in the chair. Mr Howard Marshall, the Ministry's public relations specialist, reported that it was still the feeling of the Minister that 'any interference with dogs and pets' was liable to 'cause a public outcry'. They must 'proceed with circumspection'. The zealots noted:

> The question now arises if dogs are *not* to be interfered with, they must be fed – with bread, oatmeal and milk, no doubt by their owners, all of them subsidised foods.

Mr Irving of the Ministry's Milk Division reported that

the milk consumed by cats was the equivalent of one week's human consumption. Mr Smart of Cereal Production said the flour released for dog biscuits was equivalent to two days' human consumption. Something must be done. Whatever the Minister had decided, it need not stop steps being taken to prevent there being more non-essential animals. A 'reduction by regulation' programme might just get the required results. It was therefore decided: There should be no encouragement by charities to find homes for strays, active propaganda for no more than one dog per household, no breeding of pedigrees, an outright ban on dog shows, no new dog licenses [*sic*] – and even 'penalising breeding by mongrel bitches by fines and destruction of the litter if breeding occurs'. So far there had been 'no important enforcement cases' over feeding dogs, it was noted. There had been a case in Scotland where a woman had kept fifty dogs, which she fed on meat and milk, but prosecution had been thought inadvisable. Prosecutions of pet martyrs might cause a public outcry.

But the public mind should be prepared for such a reduction by stealth – while release of any hard figures to the public would be 'dangerous' until a final policy was determined. There should be consultation, it was agreed, with the animal charities, with 'sensible' dog breeders, with dog-food makers and with 'the man in the street'.

The secret memorandum concluded:

> In regard to cats, it was the general opinion that no particular action was possible, but the possibilities of the use of *propaganda in the case of cats would be explored*.

So expect some anti-cat stories to start circulating.

But first, a pro-cat story, like this, about 'Tinker' the railway cat at Teviot Dale Station, Stockport, who had lately passed away to the great sadness of the porters. But he had not been struck by a train, the usual fate of such cats. Tinker, known as 'the biggest cat, as well as the oldest on the railways', was an exemplary ratter. 'Thousands in his time he had, and nine in a day was his best bag,' it was said.

Tinker had been on loan to Cheadle Station to do what he did best. Then he had just died. 'If you want a reason – well, there wasn't a rat left on the station to kill,' said his master. 'He kind of lost all interest in life after the last rat had gone.'

Tough on rats and mice they might be, but just as the Ministry had secretly contrived, angry letters soon began to appear in the press about cats being 'pests with which the gardener and allotment holder has to contend and guard against'. Straightaway, *The Cat* fought back.

'The Cats Protection League has always advocated keeping cats in at night,' said a stirring editorial. 'In the towns and suburbs, where the back gardener and allotment holder flourish, it is far better for the cat that it should be in at night. Do not let it be said that cat lovers stood in the path of victory for want of a little consideration and co-operation.'

That August there was plenty of anti-cat material circulating between the Ministries of Food and Agriculture, which might soon prove useful. A curious file was opened about the feline threat to the Dig for Victory campaign. It contained a growing number of cat-hating letters.

Mr G. W. Danton of Perranporth, Cornwall wrote to the Minister of Agriculture to say, 'You seem to think that the small gardeners of this country are not important enough to save their vegetables from cats.' His shallots had been scratched up twice, he complained. 'Unlike dogs, cats stroll about when and where they like.' He was sent a cyclostyled copy of *Dig for Victory News* in response, which admitted:

> Wayward cats are often very troublesome to gardeners. Owners are urged to control their pets and prevent them damaging allotments. Gardeners are advised to shoo off feline neighbours by the use of pepper dust and creating obstructions of loose wire on tops of walls.

Such advice was 'useless' according to a host of letter writers who had been sent the same missive. Outraged of Bournemouth suggested a nightly curfew on cats otherwise they should be shot on sight by the police.

Mrs G. E. Elliot wrote from Birmingham on 25 June to the Minister of Food: 'My Lord – I have a large bed of onions where I found nine or ten cats rampaging. Would I be in order in shooting or poisoning them?'

She suggested an emergency bill in Parliament for all cats to be destroyed, either that or a cat tax, and a limit of one per household. 'Having gone into the matter carefully, I am amazed,' she said, 'at the number of small households that keep large numbers of cats – ten or eleven of them. Thousands of pints of milk must be given to cats daily and the same for tins of salmon, which in fact is cheaper than cats' meat or fish.' Mrs Elliot suggested kittens be disposed of at under six weeks and an ultimatum to cat lovers be broadcast on the BBC.

A local newspaper meanwhile noted the sale for 6*d*. in Oxley, Wolverhampton, of a six-week-old kitten with 'peculiar black markings'. It was called 'Hitler'.

German cats had also better watch out. Those Nazi animal protection laws had been amended to say that cats straying outside of their usual habitat could be considered 'poaching'. They could therefore be caught but must be 'humanely' treated and their owner had to be informed. But if such adventurous cats proved ownerless, they were to be killed. Many thousands of so-called poaching cats were shot each year.

In response, Professor Dr Friedrich Schwangart, an esteemed feline researcher, had written in his 1937 book, *Vom Recht der Katze* ('On the Rights of the Cat'), that, 'domestic cats live in a permanent state of emergency, hated by the majority of the population and slandered by the press'. Third Reich cats must be protected because 'nowhere were they mistreated in any culture and country so vilely as in Germany'. Against the accusations of bird lovers that cats were bloodthirsty killers, Professor Schwangart and others presented the cat as a 'hygienic helper', eternally engaged in the unrelenting war 'against the enemies of the German people – mice'. But German cats could not look to the Führer for protection. He was Reich bird-warden of Obersalzburg, after all.

British cats meanwhile had friends in high places. During a dinner at Chequers for a US diplomat in spring 1941, a large grey cat stole into the dining room. Evidently it was Nelson, the original Admiralty cat. The American war correspondent, Quentin Reynolds, had been invited. He recorded Churchill as saying: 'Nelson is the bravest cat I ever knew. I once saw him chase a huge dog out of the Admiralty. I decided to adopt him and name him after our great naval hero.'

'His daughter, Mary, said: "You know he adopted you. He's being nice to you tonight because he knows we're having salmon for dinner and hopes you will offer him some."' It was a cosy family dinner, so Reynolds continued: 'Churchill scarcely mentioned the war. Our first course was smoked salmon and twice, when Mrs Churchill was not looking, the Prime Minister sneaked pieces of salmon to Nelson.' He should have been arrested.

There was another Prime Ministerial cat encounter on 10 August when Mr Churchill met President Franklin D. Roosevelt on board the battleship HMS *Prince of Wales*, off the coast of Newfoundland. Churchill noticed the ship's cat, 'Blackie', apparently about to desert across a gangway in favour of the cruiser USS *Augusta*, drawn up alongside. The Prime Minister bent down to pat Blackie on the head, creating a much-published news photo.

The Cat saw it and commented on Mr Churchill's *faux pas* in 'bestowing that caress abhorred by cats, rather than first offering his hand for investigation then waiting for a sign of approval'. In response the American pet press made quite a meal of *The Cat*'s remarks, calling them 'ludicrous' and 'woefully ignorant of the true nature of cats'. *The New York Times* picked it up. Mr Churchill however seemed very pleased with his meeting with Blackie,[28] beaming broadly throughout.

Back home still the tide of anti-cat invective raged. The legal advice in the Ministry of Food was that shooting, poisoning cats, etc. as angry allotment holders were

28 Blackie was renamed 'Churchill'. Later that year when the mighty battleship was sunk off Malaya by Japanese air attack with great loss of life, Churchill the cat managed to get ashore with some of the crew and ended up at Sime Road RAF Station in Singapore. 'He settled in with them, shared their rations and moved camp with them,' it was reported, 'but in February 1942, orders came to evacuate within hours and Churchill, off on one of his hunting trips, could not be found in time. Despite extensive searches, he finally had to be left to his fate.' One can hope against hope that Blackie somehow adapted to a new life in the Malayan jungle.

demanding was against the 1911 Protection of Animals Act. 'There seems to be little we can do,' noted the Permanent Under Secretary. (The internal correspondence shows a secret nest of cat lovers in the Ministry. One wrote, 'being somewhat of an admirer of pussy, I don't like the idea of classing him as a land pest'.)

Meanwhile the Soviet Army was fighting for its life as hard as Britain's cats were for theirs. On Wednesday 21 August, the Worcester and District Canine Society held a dog show in the Talbot Hotel, Worcester. Such morale-boosting events were permitted to continue when they were local affairs, raising money for a war charity perhaps. Whippet shows were popular in mining districts. No challenge certificates were at stake in this 'Sanction Show held under Kennel Club rules' – but judgements most certainly would be made. Mass-Observation was there to test the national dog-loving mood.

The show attracted a large turnout from Birmingham. Dog-friendly Silver Wings Coaches were there to take you there and back. 'All exhibitors and visitors must carry gas masks,' the police insisted.

The poet Louis MacNeice had noted in a 1938 visit to a big dog show in London how, 'hardly anywhere else can you pick up such unsconscious egotism, love me, love my dog and hate everybody else's'. In fact he was there to exhibit his own sheepdog, which was harshly judged for being far too small.

He noted how 'the toy department offers you six-foot Sapphos in breeches and hardbitten men who might be champions at billiards'. And so it seems to have been in Worcester. The Mass-Observation researcher reckoned about 150 people were there, 'overwhelmingly women, seventy per cent of them better off and mostly aged over thirty.

'Many wore riding breeches and several, including one of the judges, were in uniform, ATS and NAAFI, etc.'

He noted that several Pekingese seemed very nervous and 'would not curl their tails up'. Their owner said: 'I am surprised I didn't lose them, I thought they'd be killed.' All her windows had been blasted out in an air raid on a Birmingham suburb but the dogs had not seemed to mind.

The judge of Pekingese (Mrs Gilbert, 'Female 45 A' – gender, age, well-off, a Mass-Observation code) was heard to say to a friend: 'There is bound to be trouble, I'm just going to please myself – you can't please everybody.'

A Birmingham Dalmatian owner said it was very unfair that people were not allowed to breed thoroughbreds – the Kennel Club had asked them to breed only enough to keep going for after the war. She was always being asked for Dalmatian puppies and 'of course their whiteness is an advantage in the blackout'.

A Sealyham breeder said that she was doing very well, had recently told ten puppies and had a large store of biscuits. Small dogs were in demand. Everyone was complaining of a lack of meat. A diet of horseflesh was not keeping a large Bulldog's coat as glossy as beef, said its owner ('Female 50 A').

A homelier affair was the dog show held at Hendon Park in north London the following month in aid of the PDSA, where around 300 dogs were put through their paces. There were prizes for dogs whose owners were in Civil Defence uniform (as Air Raid Precautions had been renamed in the spring).

The Mayor of Hendon judged the tableau of child and dog entered by ARP posts and Chief Superintendent C. M. A. Steele made a speech saying it was good thing that, 'even with a war on, the people of this country could still be animal lovers and attend such shows'. In fact the show would become an annual fixture for the duration.

In November, in a bid to clamp down on profiteering, maximum prices for pets' meat were fixed: 6*d*. raw and 7½*d*. cooked on the bone. Belgian and Dutch refugees ate horse, indeed they seemed to relish it, but feeding this to pets would not be illegal. There was another variation of rationing in December when the 'points system' was introduced in addition to coupons, but which at least allowed individuals a degree of selection in getting their hands on newly restricted items such as dried fruit, cereals, treacle and tinned and bottled goods. Dog food came in tins and glass jars – middle-class dogs were very fond of it, it seemed.

The makers of Chappie dog food announced mournfully on the 16th: 'As the war goes on, a lot of us are going to wonder more and more just how we are going to solve the problem of feeding our dogs. Unfortunately "Chappie" is [now] rationed. So the most we dare promise is that customers shall get their fair and regular share of the limited supplies available.'

The Ministry of Food would take a deepening interest in Chappie and just what was in it. However much dogs liked it, the consumption of Chappie would turn out to be not in the national interest. Soon there would be no Chappie at all.

Things were looking up generally for cats, however. After much bureaucratic anguish, in early 1941, 'Peter' the Home Office cat had been granted a maintenance allowance of 1*s*. a week. The news had spread like wildfire along the corridors of power. Evacuated Ministry cats on duty in the Hydro and Hawthorne hotels in Bournemouth were clearly eligible for the same deal. 'See how elastic is the Treasury allowance for turning one cat into many cats,' suggested a senior official. It would turn out to be eminently stretchable.

A hint of better times to come came in the House of Commons on 26 November when a planted question was asked of the Junior Food Minister Major Gwilym Lloyd George – 'whether a small daily ration of milk might be provided for cats, as these animals are a national necessity?' No, said the Minister, cats consuming milk meant less for humans. But there was a glimmer of hope. 'Limited quantities of damaged dried milk no longer suitable for human food can be issued to owners of warehouses and other food stores in which cats are kept,' he added.

And there it was on 30 December when Lord Woolton, the Food Minister, announced that 'damaged milk powder' would be made available to cats – but only to cats on '*work of national importance*'.

Chapter 21
Nationally Important Cats

The Minister of Food bathed in the glow of popularity that his kind-to-cats move had bought with pet-loving public opinion. The Germans were at the gates of Moscow, Leningrad under siege and Hitler had declared war on the United States, but what really mattered at home, it seemed, was that the news for cats was good.

'Lord Woolton's face can seldom have shined with a happier benevolence than it did yesterday,' said the *Manchester Guardian* as 1942 dawned, when, 'he told the press about the largesse the New Year will bring to those engaged on work of national importance, cats who keep down mice and rats.'

But the owner of one 'extraordinarily fussy cat' told a reporter at the newspaper that her cat 'absolutely loves' powdered milk. Another woman declared her cat 'would not touch it'.

The Times editorialized on 8 January: 'We must not be jealous of cats engaged on work of national importance,' reminding readers that 'their allowance of powdered milk only extends to those engaged in keeping down mice and rats in warehouses containing at least 250 tons of food or animal feeding-stuffs.'[29]

The RSPCA hinted that the Ministry would take 'a lenient view of reasonable feeding as most homes have to keep cats to reduce the number of rats and mice' while stressing that the natural drink of cats was water, not milk. Work of national importance began at home.

There was mixed news for dogs. 'Damaged' flour and meat greaves (the residue left on a carcass after the fat has been rendered) would be released for use in dog biscuits at a level one third of the pre-war quantity, it was announced. So the turgid regime of horseflesh and oven-baked stale bread (which was technically illegal) continued.

Scraps and household waste were collected for 'economic animals' – pigs and poultry, though none for dogs. Bones similarly were part of the war effort – to make glue for aircraft and glycerine for explosives. There was still plenty of whispering against dogs as useless mouths to feed.

The Canine Defence League was still doing its campaigning best with its 'Dog Knit-for-Britain' campaign. 'Combings from almost every kind of long-haired dog, both pedigree and otherwise, have flowed into these headquarters and have been despatched to the Highland crofters,' the League had announced in the spring. And by

29 The Government's wartime approach to rats was urgent and practical. With the experience of 1940–41, when large numbers took over swathes of Coventry and Swansea after raids, regulations followed including promotion of a public force of 'rat watchers'. The Ministry of Food took over in March 1942, with a £20,000 grant to pay bounties on rats' tails. Food protection was the priority rather than public health. Professor H. R. Hewer of Imperial College, otherwise a leading expert on seals and otters, was appointed rodent control supremo.

now, 'the yarn has found its way to an industrious band of voluntary knitters who have made it into an astonishing array of "comforts"'. The secretary of the Ladies' Guild of the British Sailors' Society had written: 'I was both interested and delighted with the contents of your excellent parcel of comforts. I don't think I have ever seen anything quite like them before.'

Nor had most people. In February 1942 there was news from Edinburgh that a sock for a sea-boot had been knitted from the combings of 'Mopsy', a seven-year-old Old English Sheepdog.

The RSPCA took a wider view with the launch of its 'Help Russia's Horses' appeal the same month. The cruel masses that had oppressed the Finns were now our gallant Russian ally – whose 'cavalry operations are on a vast scale, her transport animals almost innumerable. The assault on Moscow confounded by massed regiments of the famous cossacks until at last the indomitable horsemen prevailed and compelled the first stages of the [German] retreat,' according to *The Animal World*. The appeal turned rapidly into £10,000 worth of veterinary medicines and supplies, with Mrs Winston Churchill as vice president of the War Animals' (Allies) Fund.

All things Russian were suddenly ragingly fashionable. On a winter trip to the frozen port of Polyana in 1941 a Soviet admiral presented the crew of a Royal Navy submarine with a baby reindeer. Named 'Pollyanna', naturally enough, she spent six weeks aboard. Living underwater with a reindeer posed certain challenges for the fifty-six crew. A barrel of lichen soon ran out and Pollyanna lived on scraps from the galley; she also apparently developed a taste for that wartime favourite, Carnation condensed milk. She arrived safely and was given to Regent's Park Zoo, where she lived out the rest of the war eating special 'Welsh moss' brought in by well-wishers from Radnorshire.[30]

On 8 September 1941, the Germans had closed the ring around the city of Leningrad. The Zoo, one of the northernmost in the world, had managed to evacuate some animals to Kazan. Others, such as the elephant 'Betti', died during the initial German bombardment in September 1941. However, many animals remained, among them several big cats and 'Krasavica' ('The Beautiful'), a hippopotamus.

In the years that followed until the end of the siege (January 1944), approximately a million people would die in Leningrad from hunger, cold and enemy action. Krasavica's keeper, Evdokia I. Dašina, kept the hippo alive, bringing water from the River Neva, 'even stepping into the dry pool to hug the animal and calm it down'.

That April, the RPSCA journal carried the Reuters correspondent's account of the Moscow veterinary centre, one of eight in the city, where wounded animals from the front were brought in 'special lorries'. Mr Alfirorov, regional inspector of veterinary services, showed his advanced animal hospital including an operating table for horses, sunlamps and X-ray apparatus, 'and a permanent hospital department for nursing civil horses and pets'.

Nearer the front line, the correspondent found, 'Airedales and Alsatians are used to drag wounded back

30 There were more reindeer gifts. The submarine HMS *Tigris* brought back 'Minsk' in 1941. HMS *Kent* brought one back from Murmansk, as did HMS *Belfast*, aboard which the unusual pet 'went crazy' during the action against the German battlecruiser *Scharnhorst* on 26 December 1943, and had to be shot. Its antlers reportedly adorned the cruiser's wardroom. 'Whisky', the tabby cat mascot of HMS *Duke of York*, was present (reportedly peacefully asleep) at the same action.

on little sledges; there are also a number of larger mongrels, always intelligent.'

Russomania was catching. According to the Tail-Waggers Club, their most registered dog name of 1942 was 'Timoshenko',[31] the Soviet marshal whose name was very much in the news.

The same journal recorded the work of Mrs Gardner of Hailsham, Sussex, where she had led her fellow Woman's Voluntary Service members in collecting and spinning dog hair to make service comforts. They had given a demonstration at Harrods. 'We can only use the soft combings of such dogs as Pekes, Sheepdogs, Chows, Samoyeds, Keeshonds, Poodles, Spaniels, etc.,' she wrote. 'Harsh, coarse hair will not spin. Pullovers made to our special pattern are then knitted up, and we have been able to send over sixty of these garments to the Merchant Navy with a special request that we should like them to go to the Russian Convoys.'

If the Red Army could make good use of dogs, so too could the British War Office. Mr Lloyd's experiments near Newbury had continued. After the initial plea of May 1941, the public's offer of dogs had been overwhelming. More space was needed. The Greyhound Association kennels at Northaw, near Potters Bar in Hertfordshire were requisitioned as the new 'Army War Dog Training School' in spite of what was described as 'violent opposition'. A Lieutenant Clarke would be its first commandant and Mr Lloyd its chief instructor. Even more dogs would be needed. Guard dogs, messenger dogs and 'patrol' dogs that, it was proposed, would actually

31 There were multiple Timoshenkos – cats, ponies and bears. Lots of dogs were still called 'Adolf' and 'Hitler' and there was a rash of 'Winston' Bulldogs, but there seem to have been very few pet 'Stalins'.

accompany troops into battle using their clever noses to sniff out the enemy. Nobody really knew if it would work.

Mr Lloyd was not the only war dog enthusiast. Southern England was dotted with airfields, maintenance parks and aircraft factories. They were vulnerable to sabotage and were guarded by the British Army.

Great War veteran, Major James Y. Baldwin, was another military canine enthusiast who for two decades past had been Britain's leading Alsatian breeder. On the Western Front he had come face to nose with an enemy Deutsche Schäferhunde and 'thought it was a wolf'. Entranced, he would spend two action-filled wartime years in the company of his adopted canine comrade – and in that time 'never saw an English patrol dog'.

In 1919 he and army chum, Colonel John Moore-Brabazon, the pioneer military aviator, had founded the 'Alsatian[32] Wolf Dog Club' and received Kennel Club recognition. As a reserve officer in 1940, Major Baldwin was local defence commander at Staverton Aerodrome in Gloucester, site of the Rotol airscrew factory. According to him, the army platoon guarding this vital site (they made propellers for Spitfires) was 'useless' – his Alsatians could do better.

In May 1941, his old doggy friend Moore-Brabazon became Minister of Aircraft Production. War dogs now had a powerful political patron.

On 11 September 1941, Major Baldwin staged a grand demonstration at Staverton airfield for Army and Air Force brass. They were advised to book a table at the Plough Inn in the village for luncheon afterwards. It sounded terrifically jolly.

32 They invented the suitably non-Teutonic name between them. In 1936, the 'Wolf Dog' tag would be dropped by the Kennel Club.

The whole thing was a triumph. Lt-Gen Edmond Schreiber told General Sir Alan Brooke, GOC Home Forces, how 'convincing' the plan was – 'He can get the necessary dogs given to him and trained by voluntary effort.' Sending pets to war would mean minimal cost to the Treasury. Judging by the correspondence Lt-Gen Harold Alexander, GOC-in-C, Southern Command, also caught the doggy fever.

Major Baldwin was permitted to recruit eight women and four men, 'who will be dog owners of considerable experience,' as instructors. He told the Minister, 'the thing we want the dog to do is scent any stranger within 200 yards'. By November he could report, 'everyone is dying to get at it although a very large numbers of dogs will be required'. The Treasury approved the 'experimental' dog scheme on 17 November and kennels were erected by the Royal Engineers on an AA gun site at Staverton aerodrome. It was still nominally an army operation.

Breeding would take 15 months before dogs might be useful. There would have to be a renewed public appeal for mature canines. It was Baldwin's original intention that the dogs would all be Alsatians. However it would be discovered there were not enough of them and not all wolf dog pets were wolfy enough so all sorts of dogs would have to be accepted. An epic canine saga had begun.

On 20 January 1942, the now Lt-Col Baldwin was seconded to the air force and made dog advisor and chief training officer, with an agreeable HQ at Staverton Court, a requisitioned country house near Cheltenham. The aerodrome kennels were transferred to the RAF. There was also to be a 'reception kennel' at Redditch in Worcestershire, run by a Fl-Lt Ashby, where former pets would undergo an initial two to three weeks' obedience work, assisted by the well-known Alsatian trainer, Miss M. McConnell.

A joint Army-MAP panel met on 5 March and decided all this would only work if there was a united public appeal for dogs, almost 2,500 of them. Mr Arthur Moss of the RSPCA was at the meeting, a coup on the Society's part in getting back into the Government animal business. Rations would be a matter for the War Office. A senior officer noted:

> Any dog debarred from attaching itself to a sizable cookhouse is entitled to a ration of ½ lb condemned meat and 1½ lb of biscuit – dog.

Spillers Winalot was the officially recommended rusk – if they could be obtained. Colonel Baldwin pointed out that 'tattooing the ear flap would be unpopular with owners of Alsatians as it might interfere with the carriage of the ear'. He was surely right.

The newly formed Army Veterinary and Remount Service would administer the actual getting of the dogs. At first its officers, much more accustomed to horses, seemed bemused. 'The public will be asked to lend them for the duration of the war or for as long as they are required,' it was noted at a meeting on 12 March. The RSPCA would 'register them on special cards which would then be examined and suitable dogs selected and orders issued for their despatch'. It could be assumed that the RSPCA Inspector doing so 'must know something about dogs and thus he would prevent masses of unsuitable applications'.

It was also noted that, 'if [civilian loaned dogs] are ever to be used in war, the wastage will be very high and even if they survive the war there are bound to be considerable difficulties in returning them to their owners'. It looked like a one-way ticket for pets.

And so a renewed public appeal was made in newspapers and on the wireless for dogs 'to make a valuable contribution to the national effort'. And just as the year before, the response was overwhelming. This was no longer an experiment. Now dogs were at the heart of it and would stay there. 'Over 6,000 owners have offered to lend their dogs to the Government for war service,' it would later be reported. 'The present appeal is for 200 dogs each month. The breeds required are Alsatians and crosses, Airedales, Boxers, working Collies, Bull Terriers, Kerry Blues, Labradors and curly coated Retrievers. The dogs should be not less than 10 months and not more than five years of age.'

There were all sorts of reasons to offer your pet. Mrs B. M. Harold of Norwich offered 'Tito', an eight-month-old Alsatian because, as she said, 'I would be much obliged if you could call my dog. Just at present, I find it difficult to keep him as I have two small children. Whether he turns out suitable after his trial, I still wish to have him back.'

In suburban Tolworth, southwest London, 'Khan', a five-year-old Alsatian, might be getting on a bit but seemed to be what the authorities were looking for. Eight-year-old Barry Railton, who had grown up with Khan since he was puppy, thought that the family pet should do his bit. His father – Mr Harry Railton, a clothing-shop manager – bemusedly agreed. One day, off went Khan in a van to the War Dog School. Would they ever see him again?

And so training proceeded. 'The handler continually looks after the dog, feeds it and trains it,' it was emphasized. 'Handlers are the only people who know each dog's peculiarities. Dogs must not be treated as pets by the troops.' That was very important. But how would dogs used to loving families take to this stern regime?

The first Gloucestershire trainees were sent out on guard duty that spring. On 30 April 1942, the station commander of RAF St Brides in Glamorgan reported: 'The team of eight dogs has been with us for the past ten days, handlers and dogs are in fine fettle. This method of protecting against sabotage leaves nothing to be desired.' He himself had been detected in a parked aircraft, 'while in the act of placing a time bomb'.

*

So, British dogs had a means of surviving, join up, while cats on work of national importance received an extra ration of dried milk powder. But for those pets still in Civvy Street the outlook was bleak. The Junior Food Minister, Major Gwilym Lloyd George, was asked in March whether he would consider, 'the rationing of horse-flesh and other foodstuffs for dogs and cats' – a move which would thus *guarantee they would get something*, even if it was restricted. He would not.

Was he also aware, 'that women who are working long hours complain that they cannot get suitable foodstuffs for their domestic pets?'

And they'd have to queue for hours to get it. That spring Mrs M. Clayton of Battersea, south London, declared herself to be addicted to listening to 'the rumours flying round', all of which she heard 'standing in the cat's meat line, my best source of gossip'. But what an age it took! Later in the summer she recorded in her M-O-monitored diary: 16 April 1942.

> Went over to Brixton early for the puss cat's rations. But there were terrible long queues after the Bank Holiday. I suppose I had to stand for 1½ hours.

So, what was the Junior Minister of Food going to do about it? When questioned in Parliament he replied wearily that, 'it would be rather difficult to make a census of cats'.

And as for dogs, each one was different. How could you set a ration for a Toy Poodle as opposed to a Great Dane? It was beyond the subtlest of Civil Service intellects.

'Are any steps being taken to reduce the dog population?' he was asked. 'The dog population has fallen considerably since the outbreak of war,' he replied. 'The question of food consumption by dogs is being kept under constant review.' And indeed it was.

Chapter 22
Too Many Poodles

Hunting wild animals with dogs, even in its reduced state, was hanging on by a thread. How could this be justified? In May 1942 the New Forest Buckhounds had an appeal for ration coupons turned down. 'The keepers of the forest are shooting hard,' the Lands Commissioner reported. 'Control [of deer] is better done now than for years past. In my opinion the hunt is purely maintained for sporting purposes.' Unfortunate Harriers, Beagles and Otter-hounds would also have their rations cut off altogether on 1 June.

On 22 May came news that the Sanderstead and Coulsdon Home Guard had been out with 30 guns following an invitation by the chairman of the local 'Food Production Club' to exterminate foxes that had been molesting poultry. Two cubs were claimed. The Ministry of Agriculture got excited that the Home Guard's firepower generally might be turned on foxes and hunting shut down altogether.

Next, a report arrived in the Ministry from Derbyshire that fox numbers were being kept up by the local Barlow Hunt, by means of rabbits being encouraged to breed in a special warren. They were hunting for pleasure, not to keep poultry raiders down. Local farmers were already feuding with the MFH, a Major Williams Wilson, after several similar incidents. This could be a major embarrassment.

But a much bigger hunt-related scandal was about to break. It reached the Cabinet. For months there had been rumours about a big black market operation in animal feedstuff running between the London docks, a number of barge skippers and a corrupt grain company.

Wheat, maize and barley, rationed since February 1941, had been skimmed off on a massive scale. An animal feedstuff company in Rye, in Kent had sold it on to their customers, mainly dairy farmers, without ration documents for large sums of money. Thus did black market grain turn into milk, attracting a Government subsidy via the Milk Marketing Board.

Four of the farmers were 'eminent persons': Sir William Jowitt, the Paymaster-General, Sir George Courthorpe, MP for Rye, the elderly Admiral Sir Aubrey Smith and Lord Burghley, gold medal Olympic hurdler, heir to the Marquess of Exeter, MP, guards officer, landowner and master of the East Sussex Hounds.

An anonymous letter reached Winston Churchill. 'Jowitt, your minister, is receiving stolen black market wheat. Act quickly. No patriot wants the Government to crash in scandal!' said 'a friend'.

Ministry of Food enforcement officers discreetly investigated. Jowitt claimed he had left it all to his bailiff, Mr Gough, and had 'never seen a cattle coupon in my life'. Nevertheless he had signed the cheques himself. Lord Burghley claimed innocence, telling Lord Woolton personally that it was 'unfair that every officer who is serving overseas should be prosecuted if his land agent inadvertently received too much feeding stuff'. He relied on the hunt servants to purchase quantities authorized by law, he insisted.

Sir Henry French, Permanent Under Secretary at the Ministry of Food, could see the extreme political sensitivity. He was keen to prosecute however and did 'not want a long delay between submitting the case to the Prime Minister' and criminal proceedings.

'I understand that in the case of Lord Burghley he fed it to his hounds,' he wrote, 'which makes the position *far worse*.' The others at least had turned the stolen grain into milk.

The matter was turned over to the Director of Public Prosecutions. On 20 July the Cabinet discussed it under a dummy agenda heading, 'Post War Relief'. Churchill was warned by his private secretary, Anthony Bevir: 'If there was no prosecution there might be a snarling campaign in the left wing press, who cannot like the mud being thrown at the Jews for being the arch racketeers.'

So the eminent persons did come to trial on 29 August 1942 before a special court at Canterbury on charges of 'contravening the Feeding Stuffs (Rationing) Order'. All except one pleaded guilty while claiming ignorance of their servants' actions. Lord Burghley said he had no knowledge of the matter.

'The customers of Albion Thorpe [the grain dealer] before the Court were all of the highest respectability,' *The Times* reported, 'and the prosecution had come to the conclusion that in no case did they really know or realize they were receiving excess over their coupons. They were the victims of the distributing dealers, who acted very badly.' Fines were nominal. That Lord Burghley was feeding stolen grain to his Foxhounds was not mentioned. Nor would it be. Within six months he was appointed Governor of Bermuda.

Publicising Lord Burghley's offence was just a little too sensitive in the circumstances of summer 1942. Biscuits (for humans) were put on points rationing at the beginning of August. 'Dog policy' became a renewed priority for the Ministry of Agriculture, Fisheries and Food's (MAFF) Food Utilisation Committee. An anguished political balance sheet was drafted. Cat and dog lovers might claim their pets make a valuable contribution to wartime morale but on the other hand: 'Goods have to be imported at great risk to our sailors and every ton of freight space must be freed to make room for essential supplies food and munitions.' And it was noted that the waste of food order was still untested in the matter of feeding milk to cats, but it could be. Cats might act as deterrents to mice, but if the infestation is a heavy one 'they are no longer of real use'.

The Ministry faced a conundrum – should it stir up anti-canine sentiment by revealing how much dogs were actually eating? The Minister pointed out the political perils of setting non-dog owners against dog owners. Would it really help the war effort? Educative propaganda should be considered, one dog per household be urged, maybe an increase in dog licence from 7*s*. 6*d*. to £1.

A proposal to prohibit dog shows was considered. It was pointed out that the only shows permitted were local ones, 'harmless amusement for war workers'. A conclusion was reached – 'It is not the intention of the Government to take any of the drastic repressive measures, beloved of the enemy, they prefer to leave the matter to the good sense of each man and woman who owns a cat or dog.'

But good sense could be manipulated. 'We might cause it to be known *unofficially* how much dogs do eat with a view to preparing the public mind for restriction,' minuted the Ministry's public relations expert, Howard Marshall, on 3 March 1942.

Mr Arthur Croxton Smith, chairman of the Kennel Club, told the Ministry on 6 August that they would co-operate fully in reducing of the number of dogs, but he was opposed to any policy that involved the destruction of dogs, '*other than strays*'.

The Club was fully aware, he said, 'of the political implications of interfering with existing companion dogs'. They formed ' by far the largest part of the existing canine population and no doubt consume much larger quantities of biscuit food than the show kennels do'. He was right about that.

Croxton Smith suggested a meeting of Masters of Fox Hounds, greyhound breeders and national coursing clubs to discuss the issue but it was 'not desirable that NARPAC or the animal welfare charities' should be involved. There were anti-dog plots everywhere.

Pets could not even look for spiritual comfort. That summer, Mr Leonard Noble, vice chairman of the RSPCA, wrote to William Temple, the Archbishop of Canterbury, to ask that Anglican clergy consider including a prayer for animals in their services. 'We hear of a trainload of Russian horses sent crashing into a ravine rather than let them fall into German hands,' he wrote. 'Also, we hear of thousands of lost strays and injured dogs and cats in the bombed areas.'

The Archbishop however was concerned that such a prayer might be over-sentimental and provoke ridicule. After a story appeared in *The Animal World*, the wife of a Society Inspector in Leeds sent in something suitable with a patriotic twist.

> And for those also, O Lord, the humble beasts, who with us bear the burden and heat of the day, and whose guileless lives are offered for the well-being of their countries we supplicate,
>
> Thy great tenderness of heart for Thou, Lord, shalt save both man and beast and great is Thy loving kindness, O Master, Saviour of the world.

But the Archbishop of York intervened to say it was theologically inaccurate to pray for animals, although it might be correct to pray that their treatment by humans be merciful. This particular prayer had surfaced before in 1917, supposedly from a Russian litany, but was really invented by some animal welfare society as the only way in which they could induce the clergy to pray for animals. The view prevailed. Once again wartime animals would go un-prayed for.

The Ministry of Food's clandestine anti-pet propaganda campaign began to pay off. A rash of letters to newspapers appeared. The Ministry carefully monitored them. Councillor A. R. Edwards of Moss Side, Manchester, complained about 'millions of useless cats and dogs'. London dogs alone gobbled down 180 tons of food a week, so readers of *The Times* letters page were informed on 18 June. 'Humanely exterminate all puppies and kittens at birth,' Mr A. M. Cardell of Newquay unkindly suggested on 18 July.

Mr J. Hunt Croxley, a compliant *Daily Herald* columnist, urged: 'One Dog per family! Where a husband has gone into the services, no one would begrudge the waiting woman a dog for protection and solace, but too many dog owners have too many dogs.

'And cats! No more than one cat per household should be allowed,' he continued. 'Too many cats are in the wrong places, and too many are overfed.They get milk and portions of meat and fish which to a man would be the equivalent of a three-pound joint of meat every lunchtime.'

The announcement that summer that the number of domestic backyard fowls would be restricted to one per member of the family brought a rash of letters to the Ministry about 'dogs getting biscuits that would keep my hens alive'. Canines were 'all consuming parasites'.

'Nero', a dog 'the size of a horse' – an attraction advertised at the Endcliffe Park Fair in Sheffield – caused apoplexy. 'I bet he has an appetite like one,' said a local poultry keeper, who sent a cutting about the otherwise harmless St Bernard to the Ministry. 'People are disgusted with a government that permits such abuses.' Poor Nero had no say in the matter. He was probably feeling quite hungry.

Meanwhile the RSPCA reported that they had expected to see a rise in 'deficiency diseases' in domestic animals from poor diet. In fact the opposite turned out to be the case: the variation in diet for both dogs and cats was proving healthier all round. Busy war-workers, however, had insufficient time to exercise dogs – a general concern.

In a pro-dog propaganda counterblast, the NCDL chairman, Mr Charles R. Johns, publicised the results of a rat-catching competition won by a dog living in a large town that 'scored 960 authentic kills'. The price of working Terriers had doubled, it was reported, because of the rural plague of rats – 'while sporting dogs were sought by harassed gamekeepers for estates overrun with foxes and badgers'.

The redoubtable duo of the Duchess of Hamilton and Louise Lind-af-Hageby weighed in against the 'silly attacks in the press on dogs as luxury pets'. They proclaimed:

> The value of dogs cannot be measured merely in terms of utility. They are friends of the soldiers, they are protectors of the lonely, comfort for the weary, joy to the children, often the only remaining link with past days of family happiness.

The dog lover versus dog hater scrap rumbled on; played out in saloon-bar arguments and newspaper letters columns. 'The numbers of special dogs which are doing something useful are negligible,' wrote 'Night Worker', styling himself 'a lover of clean pavements and disliker of persistent, discordant barking'.

It all reached a head in genteel Cheltenham when the owner of Lassie, 'the collecting dog', pointed out the 'many useful purposes of our dogs', including guiding the blind such as himself. 'To say nothing of the many good dogs who guard important places and also lonely women whose husbands are on war-work or active service.'

'Certainly no dog should be fed on food suitable for human consumption,' said 'Owner of Lassie'. 'Such crimes as raiding the pig-bins should be severely dealt with.' As to felines: 'If cat-haters had their way and destroyed this useful animal, how would they solve the problem of excessive vermin? May our cats and dogs be kept safe from the clutches of people who do not appreciate them.'

The Cat noted a remark made by the Minister of Agriculture, Robert Hudson, when he was heckled during a meeting. 'This town has too many poodles and too many Persian cats. What about all the food they consume?' shouted 'Disgusted of Leamington Spa'.

Mr Hudson had replied, 'We are a nation fighting for freedom. I have spent a good deal of my political life trying to prevent the issue of rules and regulations.' But the feeders of animals were beginning to feel the force of the

Imperial War Museum

It was not just ladies who hurled themselves into wartime pet rescue. London Transport bus-driver Mr. Arthur Heelas found feline fame in 1940–1 as the 'fairy godfather' of the capital's bombed-out cats.

Imperial War Museum

Imperial War Museum

Imperial War Museum

Imperial War Museum

Cats under fire showed an amazing ability to survive unscathed in bombed buildings, inspiring rejoicing when they were discovered, sometimes days later, in some impossible crevice (*opposite and top*). This lucky dog meanwhile survived to make it to the Our Dumb Friends' League (Blue Cross) hospital in Victoria – where, in December 1940, a jolly Christmas party was held for bombed-out dogs.

Getty Images

Pets were never far from war at the top. Hitler's entourage was dogs-only. He was said to regard his mistress Eva Braun's Scotties 'Stasi' and 'Negus' as 'ludicrous,' although he seems fond enough of them in this Alpine encounter at the 'Eagle's Nest'. According to reports, Negus was killed in April 1945 by a grenade and Stasi was last seen alive by neighbours of the Brauns in Munich, 'wandering the streets like many other post-war strays.'

Winston Churchill in contrast was surrounded by cats throughout. Opposite (*top*), the wartime Prime Minister makes a pet of 'Blackie' aboard HMS *Prince of Wales* in mid 1941. Poor Blackie would survive the sinking of the battleship by Japanese bombers and get to shore only to be lost in the Malayan jungle. No such fate awaited 'Mary' the goat kept by Deputy Prime Minister, Clement Attlee, in his suburban north London garden along with chickens and rabbits – one of an army of wartime urban smallholders (*right*).

Imperial War Museum

Getty Images

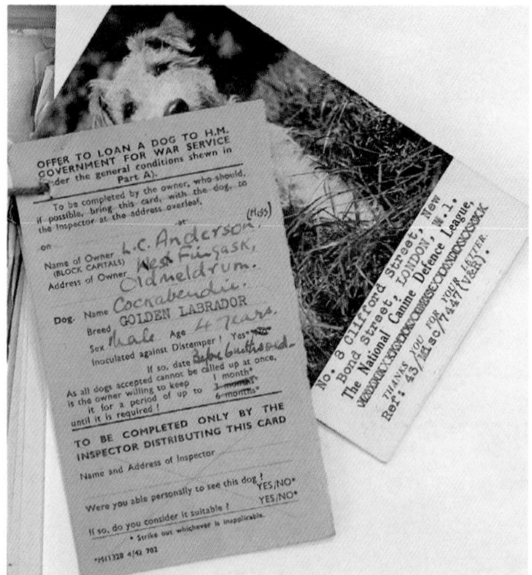

The National Archives

DOGS FOR THE ARMY

OFFERS INVITED BY WAR OFFICE

The War Office invites dog owners to lend their dogs to the Army. The breeds most suitable are Airedales, Collies (rough or smooth), Hill Collies, Crossbreds, Lurchers, and Retrievers (Labrador or Golden), although intelligence and natural ability will be the deciding factors in selection.

Dogs offered will be given an intensive course of training. Those not passing the test will be immediately returned to the owners; selected dogs will be retained for the duration of the war. Owners can have every confidence that their dogs will receive skilled care and attention.

The War Dog Training School is at Willems Barracks, Aldershot, to which offers of dogs should be made by letter.

The Times archive

With no meat ration for hungry dogs, many families responded to the call to offer their pets to the military, with tearful farewells at railway stations as they were sent off for assessment and training. Would they ever come home?

Alsatians (German Shepherds) were favoured by the RAF for airfield guard duties (*opposite top*) while the Army used various breeds for mine hunting. 'Dogs with black eyes are surly and erratic, dogs with light eyes are generally wilful,' it was noted. 'Hazel-eyed dogs' were ideal. Here (*opposite middle*) a Royal Engineer mine-dog section goes into action in Normandy soon after D-Day. Casualties were high and death in action was formally notified to grieving families by the War Office (*opposite bottom*).

Getty Images

Imperial War Museum

Imperial War Museum

Tel. No.—

Pq/660

Any further communication on this subject should be addressed to:—
The Under-Secretary of State,
The War Office
(as opposite),
and the following number quoted.

THE WAR OFFICE,
DROITWICH SPA,
WORCESTERSHIRE.

BM/V&R/632/C
Your Reference........................

14th April 1945

13 APR 1945

Sir,

It is with great regret that I write to inform you that No.3055/8488 Newfoundland X Labrador dog "Gerry" has been killed in action whilst serving with the British Armed Forces in North West Europe.

I hope that the knowledge that this brave dog served our country well, will in some measure mitigate the regret occasioned by the news of his death.

I am directed to express our appreciation of your generosity in loaning "Gerry" for War Service and our whole-hearted sympathy in your sad loss.

It is hoped that on cessation of hostilities when materials are more plentiful, certificates of service will be issued to owners in respect of all War Dogs.

I am, Sir,
Your obedient Servant,

H A Clay
Major.
for: Brigadier.
Director,
Army Veterinary & Remount Services.

J.E.Sullivan Esq.,
18 Ivorydown,
Downham, Bromley, Kent.

Imperial War Museum

Blue Cross

Time Life

There were many happy end-of-war reunions when soldiers, evacuees (and war-dogs) came home. Kennel maids prepare healthy-looking dogs at Charlton for their masters' return (*above*), while this Bristol girl is reunited with her tabby (*left*). More than seventy years later, 'service' animals are splendidly commemorated by the Animals in War memorial in Park Lane, but pets are forgotten. A substantial donation secured the inclusion of the PDSA's Dickin Medal in the monument while at the charity's pet cemetery in Ilford, east London, close to which three quarters of million ordinary pets were interred in a field, the memorial is humbler.

Christy Campbell

Christy Campbell

law. Ministry of Food enforcement officers were called to London Zoo in May after complaints that a woman fed, 'cake, Canadian apples and orange segments' to a chimp. Watching children were described as being far more in awe of what was going through the bars than the antics of those inside.

On 29 October a Mrs Isabelle Thornton, of Maidstone, Kent, was fined £15 for feeding fifty-one tins of pink salmon to cats and dogs. She claimed it was 'off' and one feels somehow certain that it was.

In Bristol, with the onset of winter, a Miss Mary O'Sullivan was fined £10 for permitting bread to be wasted. 'Her servant was twice seen throwing bread to birds in the garden,' and when Miss O'Sullivan was interviewed she admitted that bread was put out every day. 'I cannot see the birds starve,' she told the court.

And that October four company directors (one of them a serving RAF officer) were tried at the Old Bailey and sent to prison for a punitive eighteen months for 'using flour in the form of sausage rusks in the manufacture of dog food'.

London Divisional Food Office Inspector, Jane Blom, had first taken an interest in November 1941 in Messrs Shaw's Veterinary Products, the manufacturers of 'Dogjoy' and 'Livabrex' dog food.[33] Both lines were made to the same recipe – cornflake waste, fishmeal, dried liver and 'anything that they can get according to supplies' she was told. Apparently they were the same (Dogjoy was the cheaper brand sold in Woolworths). They contained no flour, so the company secretary insisted. But Inspector Blom found large quantities of mysterious sausage rusk in the Harrow Road premises plus two tons of cheese.

33 Tinned or bottled dog foods were relatively new. From none in 1937, 4,000 tons of the stuff was being sold in the UK by 1942.

Dogjoy's makers' crime was to use human-intended flour in the shape of the sausage rusk. Messrs Chappell Bros of Slough meanwhile were doing fine by Ministry inspectors who investigated their product, 'Chappie'. It was made of condemned meat, knackers' meat, potatoes and 'cereals unfit for human consumption'. Seventy per cent of it was water. That was alright by the Ministry – for now.

The year of living dangerously for Britain's cats and dogs was drawing to an end. The reconstituted NARPAC still promoted its registration scheme while drawing up baroque plans for the evacuation of coastal towns, should the threat of invasion somehow re-emerge. Clearly it had little better to do. At the evacuation end, firm instructions for the 'disposal of pets' would be issued – that no animals other than small creatures such as lapdogs would be allowed on evacuation trains.

If small animals got through to the reception end, there was no guarantee they would be allowed in billets, therefore Animal Guards 'responsible for their destruction' should operate. The Army would put down stray dogs and this time the RSPCA would be obliged to co-operate. Stray cats 'of a suitable type' able to fend for themselves would be spared and indeed would prove useful as a means of keeping down rats and mice in the empty towns. The Germans had no intentions of invading.

NARPAC had gratefully said goodbye to its rural scheme in June, the better to concentrate on pets. Economic animals were now to be the responsibility of the 'Farm Livestock Emergency Service', with the Duke of Beaufort, England's premier huntsman and owner of half of Gloucestershire, as its chairman. Fox hunting had a champion at court – Ministry officials addressed him as 'Master'.

Our Dumb Friends' League published its end-of-year report. It regretted the abandonment of NARPAC by all the animal welfare societies except itself and the PDSA as it had seen the brave Animal Guards as a 'forerunner of closer cooperation between them'. That ambition had been a little too brave.

It could report that 582 dogs and 803 cats belonging to members of HM Forces were being looked after by the League, 'in spite of those few members of the public who consider that dogs have no practical value'.

There was 'Glen', for example: 'Every month a letter is sent to his master telling him how Glen is getting on and he visits on leave.' Fifteen dogs belonging to the Fighting French brought out of France were still in their care. Meanwhile a Frenchman and his wife who had set off (with their dog) in a small boat from the Breton coast to join the Free French were picked up by a destroyer. The Setter was now happily accommodated at the Blue Cross kennels.

The perils of the Blitz were largely over. Stories of animal heroism were now of a different kind. Medals were being awarded to humans who had rescued animals (and the other way round). 'Billy', for example – a ten-month-old dog whose persistent barking woke his owners, Mr and Mrs Yerby of Kentish Town, when their house was on fire. And 'Teeny Weeny', a very brave cat who fluffed himself up to enormous size and scared off an intruder.

And to 'Jim', a 19-year-old cat in New Malden who slept downstairs but woke Mr and Mrs Coffey when he himself was woken 'apparently by smoke'. Mr Pungenti of the Old Kent Road who fell out of a tree and was killed trying to rescue a cat was specially commended (the coroner said unkindly that it was entirely his fault). The League successfully persuaded Messrs Searle's furniture shop to cancel the hire purchase agreement with the bereaved Mrs Pungenti.

The League in its end-of-1942 round-up expressed its gratitude to the Ministry of Food, 'for so readily understanding that cats are a national asset' and allocating powdered milk to the League itself for those cats it was keeping. It was concerned however that the Ministry of Agriculture had done nothing about the large amount of 'wild birds, rooks, plovers, larks, sea birds and other English wild fowl, pheasants, even swans for sale in shops and markets'.

The League was annoyed about stories that the Red Army was using dogs with explosives to destroy enemy tanks. In London, the Soviet Embassy said it was Nazi propaganda.[34]

The League was concerned too by reports of cat stealing. It talked of spasmodic but organized outbreaks of animals being stolen for their fur or even as food – as 'cat in pie or stew can taste like rabbit'. The matter was raised in the Commons but the cat pie urban myth never went away.

There were plenty of stories of unlikely animal friendships. For example, the mongrel puppy taken into a home in Weston-super-Mare, where a jealous parrot 'did its best to make the pup's life a misery'. 'But it suddenly decided to change its tune, became much friendlier and whenever the puppy was about, said: "Good morning, come right in".'

The Superintendent of the Tottenham Shelter had taken to visiting 'the local swill bins and refuse dumps' because

34 Russian dogs trained to carry satchels of explosive under enemy tanks and blow up in the process were indeed used in 1941–42 but with mixed results. German propaganda claimed it was because Soviet soldiers refused to fight.

she had found that when it was quiet, 'many stray animals congregated there'.

The League's Newport, Wales branch report had an extraordinary tale of a German aircraft brought down in Monmouthshire with a tabby cat aboard. There had been several instances of enemy pets captured at sea[35] but this was the first airborne arrival. Along with the crew it was taken prisoner. The League, in an exemplary humane gesture, offered to quarantine the tabby PoW.

'"Tiger", on arrival at the shelter, showed several characteristics of the Hun,' said the report, 'but after living under the care of the League he has become a docile, well mannered and well behaved animal.'

Nazi cats in Wales were unusual. But if Alsatians could write history, another pet-related event of 1942 would stand out. The human authority is Joseph Goebbels, the Reich Propaganda Minister, who wrote in his diary on 30 May: 'He [Hitler] has bought himself a young German Shepherd dog called "Blondi" which is the apple of his eye. He bought the dog from a minor official in the post office in Ingolstadt [in Bavaria].[36]' Goebbels continued:

> It is very nice to watch the Führer with his dog. At the moment the dog is the only living thing that is constantly with him. At night it sleeps at the foot of his bed, it is allowed into his sleeping compartment in the

35 *The Cat* had a story about a captured merchant ship with 57 German sailors landing at a Scottish port. One of them was 'a 15-year-old boy clutching a black and white cat which he guarded anxiously'.

36 The Goebbels' diaries are quite clear but other sources say that Blondi was a gift from Martin Bormann and the Ingolstadt dog was a second Schäferhund bitch, 'Bella', which Hitler subsequently kept at Berchtesgaden.

> special train and enjoys a number of privileges that no human would ever dare to claim.

There were many humans in National Socialist Germany who had no privileges whatsoever. The oft-quoted diaries of Viktor Klemperer, a Jewish journalist and academic in Dresden, published in 1998, moved many readers – especially with the account of the fate of his tomcat, 'Muschel'.

Klemperer was told that as Jews, he and his wife, Eva, could no longer make donations to the Reichsverband für das deutsche Katzenwesen (the Reich League for German cats). Its swastika-bedecked magazine, *Das deutsche Katzenwesen*, was filled with articles exalting the authentic German cat over lesser breeds. Although it was stressed in the animal journals that pets had the right to live even in times of war when feeding was difficult, this apparently did not include the pets of Jews.

In May 1942 all Jews in the Reich, and all those married to them, were told that they must surrender all pets. Dogs, cats and birds could live only in pure Aryan homes. About Muschel, Klemperer wrote:

> I feel very bitter for Eva's sake. We have so often said to each other: The tomcat's raised tail is our flag, we shall not strike it and at the victory celebrations Muschel will get a 'schnitzel from Kamm's' (the fanciest butcher here) [they had fed him on their meat rations].
>
> Unless the regime collapsed by the very next morning, we would expose the cat to an even crueller death or put me in even greater danger. (Even having him killed today is a little dangerous for me.) I left the decision to Eva.

The little animal plays, is happy and does not know it will die tomorrow. The last meal he got was veal, as in peacetime.

Eva took the animal away in the familiar cardboard cat-box to the vet in Grunaer Strasse. She was present when he was put to sleep by an anaesthetic. The cat does not suffer but *she* suffers.

As in London, so in Dresden: poor Muschel! Far, far worse was to come.

Chapter 23
The Secret War on Dogs

The surrounded German 6th Army was on the brink of surrender. If any pet was left in Stalingrad it was in immediate danger of being eaten. Fifty thousand German Army horses had perished thus far in the doomed offensive.

In January 1943, British newspaper readers were treated to the story of 'Mourka' the cat who, so an enterprising feline propagandist reported, carried messages between embattled units of Britain's gallant Soviet allies. A photograph, reportedly of Mourka, shows a fine Siberian tabby with a full winter coat perched on the edge of a foxhole in a shattered factory.

'He has shown himself worthy of Stalingrad,' reported *The Times* on the 15th, 'and whether for cat or man there can be no higher praise.' Mourka seems to have been a pet cat who somehow got swept up by the winds of war.

That same wintry day, a journalist reported seeing three cats (unaccompanied by any humans) queuing outside a fish shop in Muswell Hill, north London, waiting for it to open. Maybe they had got used to queuing after almost two-and-half years of war.[37]

The drama playing on the Volga meant Stalin could not join the discussions at Casablanca in Morocco, where Churchill and President Roosevelt were meeting to decide their next strategic moves. Mrs Clementine Churchill wrote to her husband in Africa of an urgent matter on the 14th:

> The 'Annexe' & No 10 [Downing Street] are dead and empty without you—'Smoky' wanders about disconsolate—I invite him into my room & he relieves his feelings by clawing my brocade bed-cover & when gently rebuked, biting my toe through it.[38]

Pets were not bringing much relief in Germany. The eminent military surgeon, Ferdinand Sauerbruch, recorded in his memoirs an incident in late 1942 when he was summoned to the Führer headquarters at Vinnitsa in the Ukraine. He was ushered into an empty waiting room, when suddenly, 'a huge dog entered, all teeth and snarls and prepared to spring at my throat'.

37 Fish was not rationed, partly because the Ministry of Food could not find an effective way to do so, given the underlying principle that rationed items were those of which constant supply could be guaranteed. Fish availability was governed by how many fishermen were willing to put out to sea. Queues outside fishmongers were always long.

38 'Smoky' seems to have been another grey cat, like Nelson. His entrance is not noted in Cabinet Office files. I suspect it was Mrs Churchill's name for the Admiralty adoptee, 'Nelson'. She was 'cat' or 'darling pussy cat' in their intimate letters, so confusion might be expected. 'Smokey' (*sic*) is still around in April 1955, referred to as the 'Churchill' cat in ongoing correspondence with the Treasury about a 5*s*. a week cat allowance for No. 10, but it was noted on the 18th that 'the new PM [Anthony Eden] does not like cats much and has a poodle'.

Sauerbruch then 'patted the brute, which held out a paw and began to gaze at me adoringly'. Hitler came in, in a furious mood having harangued his generals for their failures in Russia. 'What have you done with my dog?' he screamed. 'You have deprived me of the only creature who was truly faithful to me, the only creature in the whole world who loves me! I'll have him shot.' The pet (presumably Blondi) was reprieved.

Back in Britain, the war between the animal charities was as bitter as ever. For months now there had been allegations of dirty tricks. 'Do certain people want to sabotage our service?' asked the NARPAC zealot, Mr J. C. Beilby after some fund-raising skulduggery at a mayoral garden party in Blackpool. 'If they [the RSPCA] want a scrap, they can have a scrap.'

The RSPCA was now openly referred to as the 'enemy'. On 28 January 1943 the Animal Guards founder, Captain Colthurst, complained that the Society was conducting a vigorous campaign of misinformation against them, mostly by statements at meetings that the organisation was moribund and no longer had official support'.

He accused them of 'trying to cash in, by persuading our helpers to transfer their activities'. The Captain wanted 'official statements to the press and BBC [that] NARPAC is the responsible authority for the rescue and protection of animals as far as the Ministry is concerned'. This was very delicate. No excuse should be given to the 'jealous' RSPCA to attack the Ministry itself, it was noted by Home Security officials.

Five days later the Axis forces in Stalingrad surrendered. Mourka, the gallant Red Army cat, was reportedly considered for a medal, but as a civilian, was deemed ineligible.

To commemorate his African visit, the London Zoo

presented the Prime Minister with 'Rota', the Pinner lion, as a personal gift. On his return from Morocco, Mr and Mrs Churchill came to visit his new pet in Regent's Park. Rota meanwhile had fathered four cubs with his new mate, 'Janet', to be named 'Blood', 'Toil', 'Tunis' and 'Bizerta'. But it was not the end. Not even the beginning of the end for Rota's offspring.

The war on the Eastern Front had turned, but not yet Whitehall's secret war on dogs. How-to-have-fewer-canines discussions had continued since the end of 1942 as the concern of the powerful 'Lord President's Committee' chaired by Sir John Anderson, the War Cabinet's enforcer for home front social and economic policy. A supra-ministerial report was commissioned with an ostensible brief to prescribe a production figure for dog biscuits. It would end up looking for the political means for the state to kill pets.

On 19 February 1943, officials came back with a paper reporting that after the mass destruction of the first days of war, dog numbers were back to pre-war levels, four million of them, guzzling their way through 300,000 tons of food (a third of it fit for humans), which 'might be diverted to a more useful purpose'.

'With certain limited exceptions such as sheep dogs, a high percentage of the dog population makes no contribution to the war effort but is actually detrimental to it,' it was stated bluntly. All that Ministry of Food anti-dog propaganda of the year before had been in vain. The plain facts were that:

> Large amounts of material that could be usefully fed to pigs and poultry [are] being consumed by dogs. There is no doubt that bread, oatmeal, milk and other sound human foods are being illegally fed. Most of

> these products are heavily subsidised. The total amount including unrationed meat may exceed 100,000 tons, mostly bread.

The paper recommended renewed 'persistent propaganda to explain the need for a large reduction in the dog population', and 'appealing to everyone who can reasonably do so, to give up their dogs, especially in excess of one per household'. It noted however that, 'all the important dog shows have been stopped', while conceding that banning, 'small local shows [whippets] in the industrial districts of Lancashire and Yorkshire would have little effect'.

And this was new. Arrangements would be made *'for free painless destruction of those given up'*. This was a proposal for Government-subsidized dog death.

But the Committee decided on 5 March, once again, that killing pets was bad politics. Nothing should be done that would lead to 'unnecessary controversy'. There was nothing to stop a dog owner feeding his pet out of his own rations, even if it was illegal. Enforcement would need an army of snoopers. The whole thing was unworkable anyway if bread stayed unrationed. It was agreed however that the whole 'dog question would be kept under review'.

But the law on wasting food by feeding it to animals must be enforced. A Mrs King of Leamington had 'peeped into the bedroom' of her neighbours, Mr and Mrs George Fretwell of Bertle Terrace, and seen 'ten saucers of milk arranged in a ring'. They were evidently for the couple's 'fat and lazy cats', as they were described to a court.

The Fretwells had been prosecuted. The magistrate made an order for abatement of nuisance rather than 'inquire into how the milk was obtained', but wished them well with their plan to 'obtain a bungalow and large field where the cats could roam'.

Another, even more useful press report went to the Minister for attention:

> A woman at Reading was fined £2 for giving milk to a cat. The chairman of the bench asked how much milk one may give a cat. The reply was 'none whatever'.

Mr Keith Robinson of Our Dumb Friends' League was asked to comment. 'The ruling means what it says. Our animals must starve if we can't find them non-human food,' he said, 'and it is impossible to feed them properly without giving them human food. There are long queues at the horsemeat shops and usually they are sold out.' In another move, Mr Robinson accused the Ministry of Food of secretly planning 'to starve out the animal population', despite assurances from the Ministry of Home security that this was not 'desired nor intended'.

At least British housewives had something to queue for. On 25 February the author and journalist Charles Graves (brother of the poet, Robert), accompanied by his Dachshund, 'Schnitzel', went to see Arthur Croxton Smith in the Kennel Club's Mayfair headquarters to research an article on wartime dogs. He was informed that the dog population had fallen from 3.5 to 2.5 million, 'partly because owners were alarmed that dogs would go mad in air raids and bite them', and partly because the Club had restricted breeding.

Frenchmen meanwhile, so the canine grandee told him, were 'feeding their dogs on figs and powdered acorns'.

Dog food in cans and glass jars, already on points rationing, was about to be put beyond reach altogether. In the slide-rule mind of officials looking for maximum efficiency in the cause of national survival, seventy per cent of Chappie was water. What was the point of sending that all round the country? The Chappie factory at Slough could be better used for war work and the knackers' meat in it just be given direct to dogs. The company invoked the RSPCA and NCDL in their fight, without success. From April 1943, British dogs would have to go without Chappie.

Straitened dog lovers had an alternative – offer their pets to the military. While one Ministry was investigating how to get rid of dogs, the War Office still wanted pets on loan from the public while the Ministry of Aircraft Production was actually breeding them. In November the Ministry of Aircraft Production guard dog school had moved from Staverton aerodrome to Woodfold, an agreeable country house a few miles away. It was time to let the public have a glimpse. The cover of *Tail-Wagger Magazine* featured a Gloucester-bred litter of agreeable-looking Alsatian puppies, their RAF handlers, plus a goat, which presumably represented a legal source of milk.

Flt-Lt Lieutenant Hugh Bathurst-Brown, the former adjutant at Staverton, was in command with 'eight civilian trainers, the best known men and women in the country'. They included (according to *Dog World* magazine) 'Mrs. Margaret Griffin, Miss. Homan, Mr. Marshall, Mr. Dearman and Mr. Charles.' Between them, this doggy crew would one day achieve something amazing.

After visiting the top-secret training base somewhere in England, *The Dog World* could report, 'after four weeks they are far above the standard of the average pre-war dog'. And further:

> The diet is liberal and the best that could be devised. Their condition is excellent and they are as hard as iron. The dogs we saw are a hundred per cent happier than those in the average civilian home of today. The services are entirely dependent on the goodwill of the

> public in donating their dogs (on loan for the duration.) Any decent size of dog will do.

Mrs Hilda Babcock Cleaver of Allerton, Liverpool had several ideal candidates. In 1940 she had bought her first Alsatian, 'Sadie', who had had five pups. There was an advert in the paper about dogs wanted for the war effort. Colonel Baldwin liked the sound of Sadie's puppies. He was able to authorize extra rations and got authority for 'visits to local swill bins' to get the pups into shape before they left home for training at nine months old.

In April 1943 they were ready to be sent to Gloucestershire. They travelled by rail third class on an RAF travel warrant. One of them, 'Jet', would become a very famous dog indeed.

In Loughborough, in Leicestershire, at 'Glyndr', No. 6 Edelin Road, an agreeable Alsatian/Collie cross called 'Brian', bought as a pet for Miss Betty Fetch as a present from her parents, was now a bouncy full-grown dog and getting difficult to find food for. There was this appeal in the papers for dogs. With much sadness, two-year-old Brian was offered to the Army War Dog School. It was all for the best. Betty was listed as the owner from whom Brian was on loan. Brian too was set for canine fame.

In a south coast seaside town lived 'Michael', a jolly Golden Retriever, beloved of five-year-old Elizabeth Burnell. Feeding two dogs and a cat had become 'too difficult' for her doctor father. And so one day in December 1943 Michael was sent away in answer to the war dog appeal. Many years later, Elizabeth would remember saying goodbye. 'We did not know exactly what he would be doing and we did not hear much about him from then on,' she wrote. A year later there would be an extraordinary encounter:

> Quite by chance, my father was travelling to Eastbourne to visit a patient and had to change trains. As he stood waiting, a group of twelve dogs appeared on the opposite platform with their handlers and amongst the twelve, Daddy recognised Michael and quickly went over to speak to the group.
>
> The handlers said: 'You cannot talk to these dogs. They are in transit with the Army.' Daddy said: 'Yes, but that Golden Retriever happens to be mine, where are they going?' To his surprise he was told they were en route for France to search for mines.

Their lovable pet was at the sharp end of war. Would they ever meet again?

Part Four

PETS TRIUMPHANT

WE REGRET YOUR DOG HAS DIED IN ACTION

It is with deep regret that I have to inform you that (number, name, breed) was killed in action while serving with the British armed forces.

I hope that the knowledge that this brave dog was killed in the service of our country may in some sense mitigate the regret occasioned by the news of his death.

I am directed to express our appreciation of your generosity and patriotism and our whole-hearted sympathy in your sad loss.

War Office Casualty Notice, 1944

Chapter 24
Camp-followers

By 1943 Great Britain was girding itself to take the war back to the enemy on Continental Europe. There were 'foreign' soldiers everywhere. Just as East Anglia was filling with American airmen, so too the Yorkshire Wolds and the West Country would turn into vast battle-training areas. And young men far from home liked pets.

In some country areas meanwhile dogs had started mysteriously disappearing, as *The Dogs Bulletin* reported in spring 1943. Was it to make fur coats for ladies or men's gloves? Were they perhaps victims of the vivisectors? Did the Government want them for some secret purpose?

But the Canine Defenders did not have to look far. 'Dogs are found in large numbers at some camps and almost all posts have a dog or several dogs,' they discovered. 'In some camps we hear of fifty dogs as general hangers on.' Even the smallest military installation was a canine magnet with packs of playful dogs accompanying soldiers on route marches and cross-country runs.

In many cases, distressed owners never saw their missing pets again. Some were tracked down miles from home. How did they travel so far? It turned out, so the

Canine Defence League discovered, that dogs were being snatched by opportunistic dog-pedlars and being sold to soldiers from the roadside along routes taken by army trucks.

The RSPCA waded in with a plea to the Army Council that the Society should be informed beforehand when a unit got movement orders so that it might bring some humane solution to the mass of excitable pets whose playful masters had suddenly climbed into trucks and disappeared along with their cookhouse. The Duchess of Hamilton offered renewed refuge at Ferne recording the arrival of various ex-Army cats.

As well as eating all that food, there was now another reason for farmers to be grumpy about dogs. At the beginning of the war, dogs belonging to townie evacuees had been blamed (along with their owners) for general bad manners in the British countryside, rooting up crops, chasing livestock and so on. Now a huge plague of sheep worrying was being pinned on the canine camp-followers. Under pressure from the Farmers Union and the RSPCA, the Army Council posted a standing order: 'Dogs and cats in army camps must be kept within reasonable limits, dogs must be licensed and the owner have the CO's permission to keep it.'

'Stray and ownerless cats will not be abandoned, when a camp is vacated,' it said. 'Assistance of the RSPCA in their humane disposal will be sought.' Nevertheless, in the 1943 spring lambing season, there were wildly alarming stories of '10,000 killer dogs' roaming the countryside with farmers demanding they be allowed to shoot them on sight. In one night, twenty-seven of his in-lamb ewes were killed by a Collie and a Labrador, claimed an outraged Midlands sheep farmer. The position was even more serious in Scotland.

What to do with the new plague of military canines? While the RSPCA would put them down, the PDSA had an inspired idea: a humane move and publicity coup all in one. It was to make these freelance camp-followers somehow official. The Dispensary's experience in North Africa and dealing with all those desert dogs showed the way. Now there were military pets much closer to home hanging round camps all over the country.

The British Army (and others) had long had 'mascots'. Over the years a parade of bears, goats, antelopes, even a Sudanese baby boy had been formally 'attached for rations' to regiments and corps with their food and medical care provided at public expense, for as long as they 'fulfilled their role in keeping the troops amused'.

'Regimental pets' were another matter (animals defined as 'not having War Office authority and fed from government stores when civilian sources are inadequate but only on repayment'). Official or unofficial, military animals had featured in every colonial adventure and campaign, as had the Royal Navy's ship's cats.[39]

The Ministry of Agriculture however was always extremely alert to the danger of rabies (every single member of the Cairo PDSA hospital had been bitten by a rabid dog) and would rather the mother country's shores stayed unpawed by this shifting menagerie, or at least they were strictly quarantined on arrival. Recording it all therefore was a very good thing, especially if someone else was paying. Everyone could approve of mascots as long as they were qualified healthy and someone else was going to feed them.

As the Dispensary's own post-war history recorded: 'Soldiers and sailors show affection for strays. When a

39 Banned by the Ministry of Defence in 1975.

move came they became a problem. The poor, uncomprehending mascot got left behind in the deserted camp or given the key to the common.' Hampstead Heath in north London, for example – a vast tented encampment in the first years of the war – was reportedly now swarming with abandoned dogs and cats, 'all starved and in a deplorable condition'.

It was suggested that a limited number of mascots per unit be authorized, the senior officer in charge accepting responsibility for their care – 'Any animal or bird vouched for by a commissioned officer as being attached to his unit as a mascot, patrol dog, carrier pigeon or the like, can be registered as a member without any fee.' The mascots' papers would be lodged with the Imperial War Museum in London. And furthermore:

> There is a badge of membership and there is issued a medal, named after Mrs. Dickin, the founder of the P.D.S.A. In this way, animals attached to all the services are raised immeasurably in status.

The 'PDSA Allied Forces Club' would have three classes of members: the 'thousands of animals adopted by servicemen', 'animals and birds conscripted into the services for the empires' war effort' and thirdly, 'faithful beasts', honorary members of outstanding merit. Member animals must be 'serving with the armed forces or civil defence'. It was not for civilians. The first member was 'Barney' – a donkey at Hendon Air Station who had been won in a darts match.

Its original registered address was the Hampstead home of the Dispensary's master publicist Edward Bridges Webb. The secretary was Dorothea St Hill Bourne. The medal was a masterstroke and remains so down the decades.

The exotic menagerie of pets in the Mediterranean was ideal recruiting ground for the new Mascot Club. 'Bonzo' was one (a popular name, wherever) described as an 'Essex hound' who arrived one day at the PDSA Cairo hospital with two 'bronzed and tired-looking soldiers', with his eyes 'swollen, inflamed and full of sand'. Private L. A. Clarke of a 'Stores Convoy Unit of which Bonzo was mascot' enrolled him in the Club.

Private Clarke would later write in appreciation: 'I myself was lucky enough to be able to get to the hospital when Bonzo was the victim of a severe sandstorm. He was unable to open his eyes. Under the skilled hands of your representative we were able to take the dog away with his eyes open once more and after a few days of applying the ointment provided, there were no signs of trouble at all.'

But what happened next in Bonzo's war is less clear. The survival of the desert animals beyond the collapse of the Axis forces was going to be tricky. Were they to accompany their masters to new theatres of canine action? Would they find new employment in Egypt? Would they come to Britain? Some did, many did not.

'Betty' for example, a little Terrier, was found stuck in a cupboard when the Germans left Castel Benito airfield in Tripolitania in a hurry in January 1943. And 'Tocra', an 'attractive dachshund terrier cross' attached to a Panzer division, who had been captured at Benghazi in November 1942, became the mascot of the 2nd Army Signals. On the opening of the battle for the Mareth Line in March she had had five puppies. None survived, it was reported bluntly, but the mother, against all odds, reached an English quarantine kennel safely.

'Tich' was a stray mongrel bitch adopted by the King's Royal Rifle Corps in Egypt in 1941 – and accompanied the 1st Battalion all the way through the desert fighting. When the Battalion reached Algiers in 1943, the dog was placed in the care of Rifleman Thomas Walker of the carrier platoon, bringing in wounded. In early 1944 'she was smuggled aboard the ship that took the battalion to Italy' and had puppies at sea. Tich was said to be able to 'smoke a cigarette, nestling in the shade of a jeep, rolling it from one side of her mouth to the other' and to thrive on shaving water, a taste she had acquired in the desert. She would have fifteen puppies in total and survive the war. Tich too would make it all the way to England.

In great secrecy meanwhile, the first batches of Army war dogs volunteered by ordinary owners and trained at Northaw had been shipped out to the Mediterranean. These were not mascots, nor freelance regimental pets. The military vet had been right when he had predicted 'wastage' would be high. It would turn out to be massacre.

Of the dogs supplied to XIII Corps for example, the Alsatians 'Bob', 'Gyp', 'Lady' and 'Chum' in action with the 1st Surreys, one had drowned on landing in North Africa and three were killed in the bloody battle of Tebourba in Tunisia on 1 December 1942.

Alsatian bitch 'Mac' 'had presumably been captured with Major J. H. Hudson M. C. [taken prisoner in Tunisia, April 1943],' according to a terse summary of their fates. Of four dogs with the Northamptons, 'Prince', Rex', 'Toby' and 'Mark', one had burned to death, three had become ill and been shot. Plenty more one-time pets were missing, wounded and killed by enemy action.

Not all official army dogs in the Middle East were on-loan from pet lovers at home, though. According to the War Diary of the Corps of Military Police Middle East Dog Training School (formed at Almaza, Egypt, on 4 January 1943), the bulk of the dogs used for guard duties were originally donated by local civilians or were former regimental pets that had been acquired from the Italians and Germans. They would be employed in the Suez Canal Zone to combat massive pilfering of military stores by 'natives'.

By mid-May 1943 the fighting in North Africa was over and a huge bag of Axis prisoners were captured. So were a lot more animals. Of the large numbers of German military horses and mules, 'most of them, by the end, had been lost, stolen or eaten', so it would be reported. But pets, on the whole, had not gone into the cooking pot.

A dog-loving anonymous 'Flying Officer' told *Tail-Wagger Magazine* that 'an incredible number of dogs' had been attached to camps and aerodromes in the Middle East, a lot of them originally captured from the Italians.

He remembered especially, 'a tiny terrier called "Musso", which was inseparable from its young sunburned, soldier-master who promised to take the dog with him when he was posted home'. There were lots of such promises. 'He eagerly took down the name of a quarantine kennel which I gave him' said the writer of the article, who seemed to have doubts that such young men would find the resources to do such a thing in the midst of war. It was clear they were going to have to say goodbye. Were their pets just to be abandoned to the desert?

The PDSA volunteers from Cairo did what they could in traditional animal welfare style. A truck provided by GOC-in-C, British Troops in Egypt, was fitted out as a dispensary caravan with 'operating table, lethal chambers and humane killers' according to the frank and highly readable post-war history of the Dispensary by the London MP and PDSA Council member, Frederick Montague. It told these illuminating stories from the Cairo hospital:

> One soldier who was going home the following day brought three large dogs to be destroyed as he could not hand them to anyone else. He sobbed throughout and waited to see they were dead before leaving.

And there was the ATS sergeant who 'asked us to come to a nearby camp and destroy her three cats when she was also going home'. Little animal tragedies were acted out in the desert, one by one. The war was moving on.[40] Innocent pets could not follow.

40 The desert adoptee, 'Tich' (*see* p.269), was smuggled with the Battalion from North Africa to Italy, where his military career continued. There is a story of a Great Dane, called 'Beauty', the property of an RNVR sub lieutenant, who, 'drifted ashore on a Carley float and was the first dog ashore in the invasion of Sicily'.

Chapter 25
Flying Pets

All was not heart-warming goodwill in the matter of mascots as far as wartime officialdom was concerned. Rumours of airborne pets were especially worrisome to Ministry of Agriculture Inspectors. Airmen seemed to favour them as good luck charms on bombing missions.

'All these stories of flying kittens can they be believed?' so *The Cat* commented in January 1943. Unlike a ship where rats might be expected, 'the presence of a cat on an aeroplane is not necessary'. 'Aerodrome cats,' the journal conceded, 'might provide amusement and comfort to our splendid airmen. But the idea of them being taken on bombing expeditions, merely for the sake of "luck", is not an idea that commends itself.'

A picture of tousle-haired Battle of Britain pilots awaiting action in 1940 was not complete without a lovable pooch. Squadron pets were always a big feature of RAF (and now US Army Air Force) life. But whether cats or dogs, *they were meant to stay on the ground*.

There had long been rumours about animals going up in single-engine aircraft but the fleets of bombers now being assembled offered much more room for pet misbehaviour. *The Cat* would report in July that in response to 'the alleged practice of taking cats or kittens up in planes' the Air Ministry had recently pointed out – 'the carrying of dogs and cats and others by air is forbidden by Paragraph 737 of the King's Regulations and Air Council Instructions. This regulation is applicable to airmen of the Polish Air Force as well as the Royal Air Force.'

'We trust that the British example will be followed by other Allied air forces,' said *The Cat* sternly. It continued – 'The stories published about flying mascots implicate more than one of them. A newspaper, now before us, tells the story of a dog, owned by a US sergeant, which has a record of 600 hours in the air, including 50 combat flights.'

A Polish airman with the RAF, Flying Officer Stanislaw Orski, lifted the lid on what was really going on in an article for the consistently subversive *Tail-Wagger Magazine*. He laid down the 'rules' for having an airfield pet – 'first get the CO's permission, second get the President of the Mess Committee's permission to let the dog stay with you, and, third, get the kitchen staff's goodwill'. The sergeants' mess was often more reliable than the officers' mess, he advised. On flying pets he said:

> The rules say that no dogs or cats are allowed in HM aircraft but the rules are constantly being broken, plenty of pets have taken part in operations against the enemy.

But he warned that vibration and noise can cause deafness, so 'plug your dog's ears with cotton wool and wrap a heavy scarf round his head'.

Watch turning propellers on the ground, he warned, plenty of dogs have run into them. And, however amusing it may seem, 'do not let your dog run after the CO snapping at his legs. He will not find it funny.'

Some airmen were just too dashing altogether to obey pettifogging regulations. As well as 'Nigger', his well-known black Labrador, Wing Commander Guy Gibson, the CO of No. 617 Squadron, the 'Dambusters', had a cat called 'Windy' that reportedly accompanied him on combat missions – although the sources for this yarn are elusive.

'Straddle', a black Labrador, was routinely taken on Atlantic patrols in a capacious Short Sunderland flying boat of 622 Squadron RCAF, as photographs testify. The story was told of 'Peter', a Cairn Terrier who stowed away aboard the bomber aircraft of a Pilot Officer Boyd. Attacked by an enemy fighter over the target, they took damage, which showed itself on the home journey when the port-wing tank burst into flames. Peter and PO Boyd bailed out. They landed on the garden of a lady who ignored the pilot but made a great fuss of Peter.

And there was 'Antis', an Alsatian picked up as a puppy in France by refugee Czech airman Vaclav Bozděchem, who later stowed away in his master's Wellington bomber on a mission flown by No. 311 (Czech) Squadron. It has been reported that, 'Antis took part in over 30 missions over Germany – twice being wounded by flak.' In 1949 the multi-regulation-breaking Antis was awarded the Dickin Medal by Field Marshal Sir Archibald Wavell.

The record cards of the Allied Mascot Club furthermore include 'Anthony' (no. 46) a 'Black Mongrel' owned by LACW P. A. Mills, enrolled on 22 October 1943. 'Particulars of service: Flew over France and Germany on several night flights.' No further details are given.

'Flying Officer Pim' (no. 670) was evidently a hugely travelled dog. The service record of the Terrier mascot of an unidentified RAF station is remarkable:

> Accompanied his owner [Sgt J. R. Matthews] he has flown over almost every country in Europe and has about 400 flying hours to his credit. In December 1943 he baled out in Matthews's blouse. Very popular among personnel on the station.

Again there were no further details – just as well perhaps, as this was really breaking the rules. Wherever the animal originated, rabies was the fear. The seaborne Dunkirk dogs and refugee pets of 1940 from the Continent had been smartly whisked off to quarantine. They had been relatively easy to catch. So had 'Tiger', the *Luftwaffe* cat who had crash-landed in Wales to be de-Nazified by Our Dumb Friends' League. Sundry arrivals had kept coming. There had been rumours in 1942 of dogs and cats being swiped in Commando raids on the Channel Islands. Concern was raised about the cats 'which attach themselves to HM ships and which are prone to wander ashore when the ship is in port'. Animals kept arriving from all over.

'No mascot is as popular as one captured from the enemy,' commented *Tail-Wagger Magazine* quite rightly, when recounting the story of 'Peter', a German Navy dog captured in the water in February 1943 by the crew of the destroyer HMS *Montrose* after an engagement with an E-boat, off the Suffolk coast.

Our Dumb Friends' League could report, 'by 1943, the Charlton Kennels were in effect an international institution, reporting animal guests from America, Canada, Norway, Germany, France, Belgium, Holland and many other countries'. They were held in quarantine as part of the island's ramparts against rabies, with the League proudly finding the funds to do so. But with American airmen now literally flying in every day across the Atlantic or from North Africa to Britain, there was a new concern. What did these confident young men know about the Importation of Cats and Dogs Order, 1928, and would they even care? As it would turn out, they would not.

There were rumours that an American aircraft had made a forced landing at Cardinham, Cornwall, on 13 February 1943 and the crew had disappeared, along with a mysterious dog that had accompanied them, evidently from a point of origin not in Great Britain. The investigative trail was cold. Owing to operational secrecy and the dispersal of the crew it was impossible to establish where they had come from and what had become of the crew and its alleged canine member. But it was not the end of the matter.

That spring, Ministry of Agriculture sleuths discovered that 'a white Pomeranian type dog' had possibly been brought to Britain in a Flying Fortress bomber in early April. Investigators found the aircraft, Boeing B-17F 'Stella', had made a wheels-up 'pancake landing' in a ploughed field at Lychett Minster in Dorset on the 7th. It had been trying to find the airfield at St Eval in Cornwall but had overflown the landing site and run out of fuel.

Sergeant Sidney Jeans of the Dorset Constabulary had discovered a little later that 'a small white dog had been seen running about the field', which had subsequently been taken to an anti-aircraft gun site in the locality along with the crew and thence to RAF Hanworthy, the seaplane base at Poole Harbour in Dorset. After that they had gone on 'to an unknown USAAF station somewhere in this country'.

From the records, a 1st-Lt Tallmadge G. Wilson was identified as the pilot, plus eleven other crew members. They were traced in early June to Bassingbourn aerodrome in Cambridgeshire. Originally they had flown in from Morocco. The Ministry was distinctly alarmed.

Overseeing the hunt was Captain J. Fox MC, Superintending Inspector of MAFF's Animal Health Division, a qualified vet. There was plenty for him to do. Police were also investigating reports that US Army officers at the camp at Kings Weston in Avonmouth had dogs, which they admitted having smuggled into Britain. When they got there, the police could find nothing: the officers and their alleged dogs had gone.

But Captain Fox was first on the track of the mystery Pom. He would soon discover that it had indeed arrived in the Cambridgeshire fens with the Flying Fortress crew but, 'after a few hours had died in the barracks after an episode of anorexia, dullness and vomiting'. The dog had not been seen by a vet and was now buried on the edge of the aerodrome. It had not come into contact with any other dog.

Our Dumb Friends' League got involved at an official level with reports of US troops bringing in 'mascots' with no regard to quarantine. 'An outbreak of rabies would mean wholesale destruction, which would be appalling,' so Mr Keith Robinson reminded the Ministry, a menace made even more compelling because of 'the shortage of muzzles in the country'.

Then, in June, came alarming reports from near Alconbury airfield in Huntingdonshire, now a US air base. A local dog breeder, Mrs Stanley Mulcaster of Great Stukeley, had gone to the police with suspicions about 'illegally landed dogs'.

Sergeant Brookranks of the Huntingdonshire Constabulary had duly turned up at the airfield where the men of 412 Bombardment Squadron were stationed. In the barracks he had found a mongrel Chow and a Toy Terrier. The Chow ostensibly belonged to Staff Sergeant Charles F. Flynt. It was given to him by a woman in Newquay, Cornwall when they first arrived in Britain, he said.

The Terrier's apparent owner was a Sergeant Russell Matherson, who would also claim he had acquired the dog in Newquay. But another airman said it had flown in with them from French West Africa.

The baffled policeman next visited Alconbury House, a nineteenth-century pile acting as the officers' club, where he found a Lieutenant Mason with a Cocker Spaniel puppy purchased from Mrs Mulcaster, maker of the original complaint about the mystery dogs. Dogs were multiplying. The Cocker at least was undoubtedly a British dog. Lieutenant Mason could not add much.

The Sergeant and a MAFF Inspector took the matter up with the 1st Bombardment Wing Command based at Brampton Grange but on their return to Alconbury, the whole of 412 Bomb Squadron had flown, their dogs with them. Where they had gone was secret. But left behind in the base hospital they found an airman who had been injured in an accident. He told them that the Terrier was indeed his 'ship's mascot' and that Sgt Russell Matherson had won it in a poker game in Cornwall.

It was an English dog – nothing remotely French West African about it. 'The two dogs were now lodged in the Provost Marshal's quarters and appear quite healthy,' the police sergeant could report. In addition to the dogs, at least *one cat* had been landed by air from 'abroad' at Alconbury airbase, so Inspector Fox was informed.

And the plot deepened. The next day – 4 June – as well as dogs and a cat, it emerged that 'a monkey and a honey bear' had been landed from unknown points of origin, and had 'been moved off elsewhere since'. US personnel seemed to regard pet smuggling as some sort of game. Ministry officials fumed – this must be dealt with at the highest level.

It was not that Americans could not have pets; they just

had to be English ones. Second Lt Fred J. Christensen, for example, joined the 56th Fighter Group based at RAF Halesworth, Suffolk, in August 1943, flying P-47 Thunderbolt fighters. Soon to be renowned as an 'ace', the veteran pilot found a different kind of fame with the story of 'Sinbad', said to be 'a small black kitten he had found and adopted while in Britain'. No Ministry cat catcher came after Sinbad, as far as the records show.

Contemporary photographs show a lithe black cat on a pile of parachutes, and jumping on the port wing of a P-47 Thunderbolt as Lt Christensen clambers aboard. Sinbad reportedly flew, 'in the cockpit with him on many of his missions'.[41]

The lives of British pets generally were proving eventful. It might have been safer for them, like Sinbad, to keep out of the way by making combat missions over Germany. Our Dumb Friends' League reported that left-hand drive US jeeps and trucks were causing carnage among horse-drawn traffic in London.

In the capital their volunteers had conducted mass purges of feral cats in the hospital grounds of Hackney and Mile End, finding 'cases so dreadful it is quite impossible to print them'.

The Superintendent at the Chelsea Branch had been called to a flat one day that summer to 'discover that the tenant was keeping tame mice, which he allowed to run loose ... He explained tearfully that he was very fond of them and although there were over seventy, he knew each one by name. Unfortunately they had to be taken away.'

The League was also convinced 'that the abominable

41 In September 1944, when his tour of duty ended, the Suffolk cat went with Christensen to the United States to live with his family and surprised them by producing kittens.

practice of stealing cats for fur trade still continues'. That story never went away, although the 'cats-in-pies' rumours were wearing a bit thin. Also in the year-end report was the story of 'Judy', a little dog, who 'while accompanying her sailor master, had been torpedoed twice'. She had been tended at the Blue Cross kennels and 'now waits on shore with her mistress to greet her master on his leave'.

A black cat called 'Ralph' was found lying against the wall of a house. His owners had been killed nine months previously and somehow he had managed to survive. 'He would let no-one near him, but Mrs. Francis of the League's Norwood local branch, with enormous patience, tempted him out gradually with food,' said the report. 'Eventually, he gained enough confidence to come into the house and rub against her legs. He allowed himself to be stroked and from that day he did not look back.'

There were awards for humans who had shown bravery and compassion in the care of animals including, 'a Land Girl called Doris Adams, who saved two lambs from an infuriated bull'. She was awarded the League's silver medal. Quite right, too.

*

The war had turned. RAF Bomber Command was now flying deep into Germany to attack population centres by night. Animals would suffer terribly. On 24 July 1943, devastating Allied air raids destroyed three-quarters of the famous Hagenbeck Tierpark Zoo in Hamburg, killing over 700 animals.

Thus far, Berlin Zoo had been hit by a few stray bombs while a huge concrete flak-tower-cum-air-raid-shelter had risen in the nearby Tiergarten. Then on the night of 22 November, blazing 'Christmas tree' target markers dropped by Pathfinders began falling in the park as the main force followed in the darkness. The nearby UFA cinema was quickly set on fire, one tower of the eastern-pagoda style Elephant House tumbled down. '"Jenny" and "Toni", the Indian cow elephants, were standing motionless with Inra the baby elephant sleeping between them, half buried in the straw,' wrote Lutz Heck, director of the Zoo. Four others were trapped by the fallen tower now blocking the entrance to their cages.

The second wave of bombers unloaded their blast bombs and incendiaries. The roof of the Elephant House collapsed entirely, 'a curtain of fire had fallen in which the elephants and "Mtoto", the fully-grown African rhinoceros, went quickly to their doom,' wrote Heck.

The antelopes got it, the pheasantries were smashed to pieces and 'the sea-lions' basin was ringed with flames'. The Pets' Corner 'farmhouse' was set ablaze. In the big cats' house, all the leopards were dead but five 'frightened' lions came through unscathed. The dwarf hippopotamuses had been led from their blazing enclosure but repeatedly dashed back into the flames. Only one survived – found wandering, hours later, in the Tiergarten.

Two giraffes were overcome by smoke and flying glass. Monkeys escaped, chattering and whooping into the trees. 'Cleo', the orangutan, climbed up a tree and disappeared, to be found dead four days later close to the Zoo's coke heap, which had been set on fire by incendiaries. It would smoulder for a month.

The giant gorilla, 'Pongo', the 'Treasure of the Gardens', described by Heck as a 'black haired monster with blazing little eyes', escaped from his cage into the Head Keeper's house to plunge his teeth into Herr Leibetreu's leg.

There were wild stories of mass escapes; an ape and its young were seen travelling on the U-bahn, a wolf taking tea in the Eden Hotel. A diplomatic official wrote two days after the raid: 'Fantastic rumours are circulating. There are crocodiles and giant snakes lurking in the hedgerows of the Landwehr Canal.' There was a story that an escaped tiger made its way into the Café Josty on Potsdamer Platz, gobbled up a piece of pastry and promptly died.

The RAF returned the next night. The Aquarium was comprehensively wrecked, smashing the 'Forest River' and tumbling crocodiles and alligators into the sub basement, 'wounded by bomb splinters and writhing in pain'. Liberated snakes became torpid as the Berlin November cold stole into the smashed tropical tableaux. Days would follow rounding up survivors and striving to somehow keep them warm. Many animals, 'ponies, zebras, horned hogs and anthropoid apes', were lodged with well-wishers in the suburbs and countryside around.

The Herr Director was frank about the fate of the dead. 'Very good were the crocodiles' tails cooked until they were very soft,' he would write. 'The deer, buffaloes and antelopes supplied meals for humans and animals. Later bear bacon and bear sausages became a particular delicacy for us.'

Chapter 26
The D-Day Dogs

Britain's dogs had seen off the threat of state-sponsored destruction. Their numbers were up and their tails were wagging. The perils now faced by pets in this fifth year of war, 1944, were the usual ones of casual cruelty, abandonment and intermittent food supply. Feral packs of cats and dogs roamed urban bomb-sites and the fringes of military encampments, which bloomed across the country from end to end. The pickings were good.

But one pampered pet was in trouble. The Metropolitan Police intervened when a Wire Haired Fox Terrier was reported for 'assaulting ladies in the street'. It happened to belong to General Martin Valian, commander of the Free French Air Force. A Foreign Office memo recorded a series of springtime 1944 incidents in London – when the dog 'bit one lady on the right thigh and tore a considerable rent in the coat of another lady'. The fate thereafter of the General's dog was not recorded.

A hidden drama was about to begin at London Zoo. Its year-long course would be lovingly reported by the *Evening Standard*'s Zoo Correspondent. In the early spring a tabby cat who had evidently been living wild in Regent's Park had somehow got into the depleted 'reptiliary' and found an empty 'hidey hole' where venomous snakes once basked and gave birth to five kittens. She had another litter close to the piggery. The keepers called her 'Sally'. But what to do with the cat and her multiplying offspring? Would they be a threat to the other creatures, the exotic birds or small mammals perhaps? Sally, as it turned out, would have her own ideas.

The so-called 'Baby Blitz', the *Luftwaffe* air attacks on southern England between January and May, did not affect pets much. What pet owners could not know (but British Intelligence did) was that the Germans were planning a renewed aerial assault on the capital using 'flying bombs' launched from northern France ('V1s') and rockets ('V2s') of as yet undetermined destructive power. They might contain gas or worse. Plans were being made anew for the wholesale evacuation of London.

The summer 1943 saga of the flying pets had meanwhile developed multiple new sub-plots. In November, three US Army Air Force NCOs had been dedicated to 'special duty in detecting illegally imported animals'. The smuggling of cats and dogs had become a matter for the 'Chief Veterinarian, ETOUSA' (American Army, European Theater of Operations), Colonel Edward M. Curley.

Miscreant officers and their pets were being tracked down. Examples must be made. It emerged that in January 1944 a certain Captain A. M. Russell arrived by air with a dog at St Mawgan in Cornwall and had been detected by a US Army vet. Eighth Air Force HQ was notified, as was the Ministry of Agriculture, but dog and owner had since disappeared into the fog of war. Five months of hunting ended when the Captain was found – to say the dog in question had been accidentally killed a month before.

The US dog catchers could also report meanwhile that a Pekingese had been smuggled ashore from a Cunard liner troopship at Liverpool in April. More dogs arrived in the north-western port aboard the SS *Mauretania*. They had all disappeared.

The Agriculture Minister was asked in the Commons about 'a dog belonging to a Flying Fortress crew, which was spared on the personal intervention of President Roosevelt'.

'Everything we see in the Press is not necessarily correct,' the Minister replied. The bizarre story was all over the US papers and had been picked up in London. 'Rob Roy', a Cocker Spaniel, had been seized on arrival after a transatlantic flight by the British authorities from B-17 Bombardier Lt Jack Roberts. He had written to his mother in Atlanta, Georgia, to tell her that his pet 'might be executed'.

Mrs Roberts had appealed to the President. After all, Rob Roy was an all-American dog. Whatever the diplomatic trail of events, it turned out that the Cocker was now in quarantine in Croydon under the care of Captain F. J. Richmond, who promised the pet was safe so long as its board and keep were paid for. 'Telek', General Dwight D. Eisenhower's Scottie acquired in North Africa, was in the same pound, the British Army vet pointed out.[42]

One dog that slipped through undetected was 'Recon', a Jack Russell Terrier, who was obtained by a Sgt Nelson at Boise Air Force Base, Idaho. Having recovered from a rattlesnake bite, Recon became something of a squadron hero. According to the unit history: 'Recon went through all of the 1942 stateside training, was smuggled aboard the *Queen Mary* for the trip to England, was a faithful 427th

42 Telek's name was reportedly a combination of 'Telegraph Cottage', a substantial suburban villa in Kingston, south-west London where he was quartered and Kay Summersby, the Supreme Commander's twenty-six-year-old driver, with whom he was having an affair.

Bomber Squadron mascot at Molesworth [Cambridgeshire] and was transferred to North Africa.' The unfortunate dog was promptly 'killed in a Jeep accident in Casablanca'.

Faced by such shenanigans, Sir Daniel Cabot, Chief Vet of the Ministry of Agriculture, despaired of tracking down any American dogs. He suggested to Col Curley that 'one of his officers be entrusted with the task'.[43]

This was getting political: flying dogs had reached the White House. A major diplomatic incident involving animals would be most unfortunate. Colonel Curley appointed Lt-Col Benjamin D. Blood of the US Army Veterinary Corps to bring some rigour to the proceedings.

Lt-Col Blood could report on 19 July that fifty-nine dogs had been illegally imported by US Army Air Force personnel, of which thirty-seven had been destroyed and nineteen placed in quarantine. In addition there were nine monkeys, two parrots and one honey bear, all of which had been reported to the authorities and disposed of in accordance with British law.

One dog was missing. Another had died on landing and one was missing in action, having been 'taken on a bombing mission'.

Some English pets were meanwhile being flown into action officially, at least in training. 'Brian', for example, Miss Bettie Fetch's Alsatian-Collie cross from Loughborough, had passed out of Northaw in early 1944 as a patrol dog and been chosen to go to the 13th Parachute Battalion for special training at Larkhill Camp in Wiltshire.

He would be a member of the unit's scout and sniper platoon. In charge was Lance Cpl Kenneth Bailey, formerly

43 The Veterinary Corps ranking officer indicated to the MAFF at US Strategic Air Forces Europe, High Wycombe, was a Major Max G. Badger.

of the Royal Army Veterinary Corps. There were three more dogs: 'Bing', lent by the Cory family of Jackson Avenue, Rochester, and another dog, name unknown. War Dog 'Glen', origin unknown, was assigned to A. Coy, 9th Parachute Battalion.

The dogs' jump training began in early April. Lance Cpl Bailey recalled: 'I carried with me the dogs' feed consisting of a two-pound piece of meat, and the dog was readily aware of this.'

Of Glen it was said, 'he loved to jump', equipped with a parachute harness and a little red light on his back, and was trained to stand still as soon as he hit the ground. 'Everybody loved him, he was the pet of the battalion, but no one was allowed to pet or feed him, say good dog or anything like that.'

Brian jumped readily on the green light and 'wagged his tail vigorously' during the descent. According to his trainer: 'The dog touched down completely relaxed, making no attempt to anticipate or resist the landing, rolled over once, scrambled to his feet and stood looking round.' Each dog made four descents, after which they resumed a normal existence. Then came D-Day.

At fifty minutes after midnight on 6 June, the dogs took off from airfields in Oxfordshire. Brian and Bing, and a third dog, descended by parachute around the village of Ranville; Brian got hung up in a tree and was shot at. He was rescued by his handler and reported to his post on the edge of the Bois de Bavent – 'He subsequently endured heavy mortar and shellfire, during which he was slightly wounded, but, with the provision of his own slit trench, survived.' Bing was wounded; the third dog disappeared.

On the approach to the Merville jump, Glen was terrified by the ground-fire and had to be dragged out from under the seat and physically thrown out of the transport aircraft door. The drop was way off target. Glen and his handler, nineteen-year-old Pte Emil Corteil, managed to rendezvous with their brigade commander, Brigadier James Hill, at the village of Varaville in the early hours of 6 June.

But on the trek towards the objective, Pte Corteil and Glen were killed, along with many others, by a disastrously mislaid stick of RAF bombs. They were buried together on the insistence of Major Parry, Corteil's company commander, who had led the assault force against the Merville Battery the night before. Parry believed that since they were so devoted to each other in life, it was proper that they should share the same grave.[44] Glen was once someone's pet. Like his young master, he would not be returning to a loving family.

Two dogs that went to France were 'Scruffy' and 'Knocker', who crossed the Channel with an RASC service unit not long after D-Day. They seem to have been smuggled. One of their number recorded, 'I got over there a few days later and there was Scruffy, hanging round the kitchen as usual.' But where was Knocker? 'I was told he had been killed by enemy fire two days after getting to France. I found Scruffy scrabbling by Knocker's grave,' he wrote. 'Was he burying a bone? I don't know. I left him there.'

*

D-Day had opened a second front in the war on smuggled pets. Allied forces were ashore in strength in Normandy.

44 They are buried in the Ranville War Cemetery in Normandy, France. Glen is the only animal to be buried in a Commonwealth War Grave. It bears an epitaph written by Pte Corteil's mother: 'Had you known our boy you would have loved him too. "Glen", his paratroop dog, was killed with him.'

Prisoners were being shipped back across the Channel. Would seaborne pets be coming with them? On 2 July the US guards of the extemporized Prisoner of War camp at Portland, Dorset, handed over to baffled local police a 'black and tan Alsatian bitch'.

Was it a pet, a mascot, or a Prisoner of War? It seemed to be the latter for it had a US Medical Department label:

> Name: Harry Hollmer
>
> Line of duty: G.P.W. [German prisoner of war]
>
> Location where tagged: D-V-POW A.P.O. 155 [The US Army field post office, Dorchester]

Was it subject to the Geneva Convention? Was it a French civilian animal? Now it was a 'stray' on police hands and in the same perilous boat as many thousands of abandoned British pets. Technically it had three days to live.

That night the female prisoner had a litter of puppies. On being informed of the development, the Chief Constable in Dorchester told the Ministry of Agriculture in London, who told Col Curley. The dog had been landed from LST 520 (a tank-landing ship) of the US Navy in the possession of a prisoner listed as 'unknown, ex-Normandy beachhead'. It was the third such dog brought back from France. Two of its predecessors had been destroyed, so the police message for the Ministry reported. Would this one go the same way?

For a further twenty-five days, mother and pups awaited their fate at Portland police station. It was referred back to the Americans and reference made to the mysterious 'Harry Hollmer', who had accompanied the dog. Was he an American GI or a German prisoner? With the quarantine option still open for mother, offspring, or the whole family, it would be necessary to know who would pay, so Mr S. P. Maddison of the Ministry's Animal Health Branch told Colonel Curley.

But the Colonel politely insisted that, with their handover of the dog to the police, it was now a British problem. On 27 July the Ministry's Dog-Finder-General, Captain Fox, went to see the prisoners. The pups were now almost a month old. 'Retention is becoming irksome and not without a certain amount of danger,' he reported. 'The dog appears healthy but is restless. It should either be destroyed or removed to approve quarantine kennels.'

From the file it is clear that some Ministry officials were trying their best to save the D-Day dogs. It was suggested they be offered to a Major M.A. Murphy, RAVC, head vet of Eastern Command, to see if the War Office would take them. His superior, Major Bridgeman, was keen.

He contacted the War Dogs Training School. The report of what happened next is terse. Major D. Danby, chief vet at the school, telephoned the police at Portland and told them it was not a British Army dog that had somehow been waylaid. The Commandant, Major Bridgeman, decided that if they took them, it would somehow 'impinge on the French authorities' right to ownership.' It went up to the 'Brigadier responsible for war dogs (Brig. C. A. Murray, Director Army Veterinary and Remount Services),' who agreed.

There was to be no new home in leafy Hertfordshire for mother and pups. The report reads:

> The decision to tell the local authorities that the Ministry would not object to the destruction of the bitch and her puppies was telephoned to Mrs. Fox for communication to Captain Fox.

The death notice was somewhat terse:

> Destroyed by shooting on 1 August and carcass cremated at local gas works. Three pups destroyed by gassing and similarly disposed of.

There were luckier D-Day dogs. 'Fritz', a St Bernard, was captured by men of the Hampshire Regiment on D+1, 7 June, at Arromanches in Normandy. 'Somehow or other he got aboard the landing craft while the prisoners were embarking for Southampton, where on his arrival he presented a knotty problem because of quarantine regulations,' said a near-contemporary account.

His Allied Mascot Club card (no. 439) reads: 'Might have been condemned to death on arriving in this country had it not been for the kindness of Leading Wren Elgar who offered to pay his quarantine and also presented Fritz to the regiment that captured him. He appears at all the regimental parades and official records are kept of him like a soldier.'

Another D-Day dog was 'Sailor', a St Bernard spotted running on the sands of Gold Beach at La Rivière on 6 June. According to an account in *Tail-Wagger Magazine* that summer, Able Seaman Curtis stayed on the beach for five days with the dog in a shelter, before taking him back to England on a warship. They were met at a 'West Country port' (almost certainly Weymouth) and the dog's fate became a matter for the authorities. But Mr C. E. Dowdeswell, secretary of the local branch of the RSPCA, took him on and now 'Sailor' was being offered to the public when his quarantine was complete by the 'Dogs of Britain Red Cross Appeal', care of the Kennel Club. Lucky Sailor!

There was a second dog called 'Fritz', an Irish Setter belonging to General Bernard-Hermann Ramcke, German commander of the besieged Brest fortress, captured on its surrender in September. The event was filmed by US Army combat cameramen. Glossy Fritz is on a lead and seems to obey every command of his combat-jacketed Teutonic master.

According to one source, General Ramcke had been promised by Gen Middleton, CO of the US 8th Infantry Division, that he and his dog would not be separated. But when master and hound were delivered to a certain Major Becker of the British Army at an airfield in southern England, Becker quite rightly announced, 'Herr General, the dog has not been tested for rabies!'

Ramcke was outraged but Major Becker refused to be cowed. He interrupted the General's tirade with, 'You are in England now!' Gesturing towards the dog, the Major ordered: 'Take him away!' according to the memoirs of his US military police guard. *PDSA News* reported simply, 'the dog has been flown to the UK and sold to an Englishman and taught to understand and obey commands in English'. Lucky Fritz!

Chapter 27
Doodlebug Summer

A week after the D-Day landings, the *Luftwaffe* had opened the bombardment of London with flying bombs, soon to be popularly called 'doodlebugs'. Although unheralded this time by Government announcement or by 1939-style panic, there was another wave of pet killing. One south London woman recorded standing in 'a long, long line of white-faced women to have my pet destroyed because the kennels were blasted out'.

'Ronnie', the pet cat of an Uxbridge family, 'returned home the afternoon of the next day' after their house was shattered by a flying bomb. He was completely black. 'We had been distressed at losing him, but we took him to the veterinary surgeon and he was put painlessly to sleep,' said Ronnie's owner. 'After all, we were having to depend on friends for a night's shelter ourselves and Ronnie had led a very happy life,' she added.

The forewarned defenders achieved some success but enough missiles got through to make the doodlebug summer (9 June to 1 September 1944) extremely uncomfortable for London pets. Now the rituals of evacuating children and hastening to the Anderson shelter in the garden, or Morrison under the kitchen table (unused for the past three years), were re-enacted. This time, many flying bombs fell in the suburbs, crashed, or were shot down in open fields in Kent and Sussex.

The London Zoo was hit on 27–28 July. A flying bomb fell in Regent's Canal while another blew up in trees in the middle of the Zoological Gardens killing two sulphur-crested cockatoos, a Silver Pheasant and a Sonnerat's jungle fowl. Later a fatuous newspaper row would blow up over priority given to repair of the komodo dragons' den while houses in Prince Albert Road still needed the roofs to be repaired.

NARPAC barely functioned in this new emergency (it would wind itself up at the end of the year). In a report for the Minister, Mr Edward Snelling, the long-suffering Ministry of Home Security official, noted that for 'some time past relations between the constituent members, the veterinary profession, Our Dumb Friends' League and the PDSA had been so strained that they had in effect been working independently'. He thought that Edward Bridges Webb deserved his 'scoop' in grabbing the registration operation and outfoxing the veterinary profession, who still regarded the People's Dispensary as 'quacks'.

It was the charities who were in the front line again, just as they had effectively been all along. The Canine Defence League's *The Dogs Bulletin* proudly announced:

> None of our shelters were hit in 1940–1 but now three clinics have been destroyed by V1s. Londoners with their canine and feline friends will stand fast!

PDSA News reported their efforts under the doodlebug assault in similar terms to the Blitz – with lots of putting to sleep of wounded animals in the ruins. In August 1944, 'Spot' received the Dickin Medal (as a self-starting rescue dog) for digging himself out of a blasted house and barking for rescuers to come. He himself scrabbled and dug at the ruins. 'Their faithful little friend with four bloodied paws is being treated by the PDSA and is now almost recovered,' the journal reported in September.

The Dumb Friends' League shelter at Hammersmith was shattered by a flying bomb and the Wandsworth branch at 82 Garratt Lane (now under a giant Sainsbury's) was entirely demolished. 'Fortunately the animals were mainly at the back of the building and suffered little hurt other than shock,' the 1944 report stated. It also recorded a swan injured by a flying bomb, which was taken to the master of the vintner's company – 'The bird was put to sleep.'

The League's Blue Cross medal[45] was awarded to 'Pussy Wake', 'who saved his family when a fire broke out in a downstairs room. He ran upstairs and awakened them by scratching at the door where they were sleeping.' 'Rex' was another recipient, who, 'when he heard a flying bomb approaching, dashed up the stairs to the room of nineteen year old Rosene Mason, warning her in time for her to reach safety before her room was wrecked'.

'Ruff' received a B. C. 'When a flying bomb wrecked his home, he attracted the attention of the rescue party to the debris, under which his mistress and her baby were lying trapped. He actually gripped the nightdress of the infant, pulling the child to safety.'

'Rex', a Retriever, saved the life of women in East India Dock Road, trapped by the explosion of a flying bomb. Then there was 'Whiskey', a cat who 'saved Corporal

45 The Blue Cross medal, originating in 1906 for humans and from 1940 for animal bravery, never quite achieved the same kudos of the subsequently far better known Dickin Medal of the PDSA. In 1945, the RSPCA instigated the red collar and medallion 'For Valour'.

Witcomb's family when fire broke out in the house'. Heroes all!

In the original Blitz, animals had not been allowed into London County Council rest centres. By summer 1944 they were permitted to do so provided they 'were under control and not a nuisance to others'. Meanwhile public air raid shelters remained barred to pets. A lot of Londoners chose to see it through in the backyard Anderson shelter or the indoor Morrison (a kind of steel cage that could serve as a kitchen table, named after the Home Secretary, Herbert Morrison). The RSPCA offered sound advice 'to make an improvised dog shelter in the home, and train them to go there on command'.

'I have four such dog shelters in my narrow hall for my Charlies [King Charles Spaniels],' wrote a correspondent for *The Animal World* in late summer. 'I find my dogs will go hurriedly in during the daytime when told to Shelter! Shelter! We throw our bed covers and overcoats over the dog boxes at night for extra protection.'

She also suggested, 'taking in some stranger with an animal'. In a very humane account, the anonymous writer told how: 'In my last home, Wardens used to send me solitary old women who had only the one dog in the world. They refused to be parted. They could not go to the shelter with the dog. I remember one who said would I take her dog for the night, and her face lit up with joy when I said – "I can take no dogs without their owners".' She noted however that 'only a miracle can save you and your creatures' from a direct hit by a flying bomb.

There were plenty of sentimental tales of human-animal bonding to rival the dramas of four years before. One old man was seen by an RSPCA Inspector to go first into his garden and clear a space on the ground, then to enter the wreckage of his home and bring out 'two dead cats, one black, and one tabby'. The inspector approached the old man, who said: 'I have lost everything. My two poor cats were killed outright, but the least I can do is give them a decent burial.'

Teenager Harry Atterbury recalled his pet adventure in a short memoir. Having returned from evacuation to the embattled capital, on a quiet Sunday morning, a V1 crashed at the corner of his home in a street in Islington, north London. 'My parents' home was demolished with me underneath. I was pulled out,' he wrote.

> Over the next few weeks, I returned to dig in the debris of our home although there was little left to recover. But two weeks later, when raising up the corner of our flattened kitchen table, I was startled when a lump of fur moved and I took into my hands our old pet cat, still alive. On the bus to Stoke Newington so many other passengers expressed sympathy for her, even in that filthy state. She lived with us for several years after.

And as in 1940–41, there were tale of cats returning to the ruins and benign interventions by strangers. *The Evening Standard* reported the story of Mrs A. Emery, a 'daily help' and widow of a Royal Marine 'who for five weeks has given her milk ration to over 20 cats left homeless by flying-bomb attacks'. She said:

> I have been bombed out twice and lost my own cat, Timmie, who was 10 years old. Because he meant so much to my little household, I could not see other cats waiting near houses where people had been killed or evacuated, without feeding them.

Another woman was seen to take a saucer and a pint bottle of milk from a shopping bag, according to the same report. 'She placed a saucer of milk before "Major", a ginger cat, reputed by the police to have killed over fifty rats and mice in a week. Major's owner, a school teacher, was killed in a recent flying bomb attack, but the cat has since kept constant watch outside the ruins of her house.'

Thirty years after the doodlebug drama, the social historian Norman Longmate recorded first-hand (from owners) accounts of how their animals reacted. 'Our dog soon realized that the sound [of an approaching V1] might well culminate in bang and appeared to listen anxiously,' found a woman living in Kent. 'The cat appeared to take no notice whatever.'

'Kim', a Bedlington Terrier of Shirley, Surrey 'used to get most distressed,' her owner remembered, 'when we just had to carry on with our lives and ignore the doodlebugs until the moment of cut-out.'

'Mickie', a Shepherd's Bush Collie, 'was usually first into the Anderson shelter [and] put on a great display of courage once the danger was past, dashing out the moment the all clear sounded and barking boastfully.' An unnamed Scottish Terrier in Neasden was 'always first into the Morrison [shelter] and the last to leave until the dog got quite confused and started going into the shelter on the all clear'.

Just as in the Blitz, pets were alerted to incoming danger before human perception. 'Dawn', an Eastbourne Elkhound, would 'prick up her ears, emit a shrill bark' and then, 'her duty done, make for safety under the stairs'. 'Benjamin', a Springer Spaniel in Sevenoaks, 'would suddenly wake, stand in the middle of the room and "point" – this gave his owners time to get to the shelter'.

The *Daily Mail* featured the 'Achtung Chimps' of London Zoo, who could 'hear the flying bombs miles away'. When the alert first sounded they carried on as normal, but – 'then whimpered, let out a series of yelps and retreated to the furthest corner of their cages'. Once they had heard the engine cut out and the rumble of the explosion, the chimps returned to their usual antics.

Cats also gave warning of approaching danger. 'Binkie', a Sussex cat, would dive for cover seconds before the oncoming V1's pulse-jet motor could be heard by humans. And there was 'Sandy', the Eastbourne cat, who one day 'refused to come out of the cupboard under the stairs after a V1 had passed'. Sandy's owner then looked outside, 'only to see another doodlebug almost overhead'. Its engine had stopped and it was about to dive.

The Cat told the story of 'Ticky', a 'timid cat' who found the courage to dash into danger and rescue her kitten, dragging it into a garden Anderson shelter as flying bombs droned overhead. Nervous Ticky thereafter would sit on the garden wall scanning the sky and, as her owner wrote, 'it is pointless even to duck if Ticky has not moved from her observation post. She never mistakes the direction of the sound and her powers of perception are much better than mine.'

Caged birds often died outright of shock when a V1 impacted, while backyard food animals were routinely blasted. *The Animal World* reported, 'one street in north London in which all the human inhabitants had been killed or severely injured but their backyard fowls were untouched. Emergency feeding of a mass of hungry chickens was required.'

In the open fields meanwhile, 'more cattle were killed by blast than flying splinters in V1 raids', according to a Farm Livestock Emergency Service (FLES) bulletin, while a number of hapless cows died, 'as a result of eating pieces of wire left lying around after explosions of V bombs'.

A Kent Land Girl recalled for the social historian Norman Longmate how the cows in her charge were 'petrified' by the first V1s and tugged at the chains tethering them, but soon went back to their former placidity. Horses proved as impassive as cows. Rural hens were surprisingly resilient.

Useful quantities of meat could be recovered from blasted herds. 'When a flying bomb dropped among thirty-two pedigree Friesians, the slaughterhouse manager and veterinary surgeon attended the incident,' an FLES report stated. 'Highly satisfactory salvage of seven cattle resulted, 1,748 lb of meat were recovered and the nine cattle injured were all treated. Fifteen animals were unaffected.'

A week after the last flying bomb was launched from northern France, the first V2 rocket to hit London arced across the North Sea, fired from Holland. This was a different kind of war. It travelled at four times the speed of sound. For weeks to come the Government would claim the strange double bangs rocking the capital were 'gas main explosions'. Nor would the Germans admit what they were. Animal instinct clearly knew better and there are plenty of stories of psychic premonitions.

The very first rocket hit Chiswick in the southwest of the capital on the evening of 8 September. An engineering worker's wife was alone in a house in Wilmington Avenue, about 300 yards from the impact. Home from work, she had put down a bowl of food for her cat, 'Billy' – 'large and white and not particularly friendly'. Billy ignored supper and, unusually for him, tried to jump onto his owner's lap. She recalled:

> I stood up and lifted him up in my arms. He was quite still for a second then suddenly he leapt from my arms

> and rushed out and through the cellar door, which was just outside. Before I could gather my senses there was an almighty explosion. The cat would not come up out of cellar, not even for his food.

A nurse living in Balham observed the family cat, 'Junior', asleep on a chair. Suddenly he jumped up in the air, 'gave a wail of terror' and rushed under the sideboard in the corner of the room, just before the 'terrifying explosion' of the first local V2. Junior was to show the same prescient power on two more occasions. Quite soon pets themselves would intervene in the 'revenge weapon' episode in a quite remarkable way (*see* p.308–14).

There was a happier event in the rocket-blasted autumn of 1944, but one still shrouded in the deepest mystery. Our Dumb Friends' League reported: 'At the request of the Palace authorities, the League collected two kittens that had been born at Buckingham Palace. They were found good homes.'

The Royal Family, as far as is known, had had no cats since the time of Queen Alexandra (who adored Siamese). But there were plenty of dogs. Since the thirties, the favoured royal breed had been Corgis, starting with Golden Eagle, known as 'Dookie', a Pembrokeshire acquired in 1933 from the Rozavel Kennels in Betchworth, Surrey, of the breeder and pre-war supplier of pets to top Nazis (*see* p.116), Mrs Thelma Gray.

Dookie died shortly after the outbreak of war and a second dog, 'Rozavel Lady Jane', was run over in Windsor Great Park at Christmas 1943. As a replacement for 'these two little oddities' (as *The Times* had unkindly called them on their first public appearance), Princess Elizabeth was given her own Corgi for her 18th birthday on 21 April 1944. The two-month-old puppy was named 'Susan'.

So where did these mystery kittens come from? The Palace has stayed silent for seventy years. At the time the *Daily Mirror* sniffed out the story and reported on 17 November: 'Puzzle of the Palace Kittens. "Belinda" and "Jane", born in the Silver Room of the palace, have lived in the royal presence for ten weeks before being given to the Our Dumb Friends' League ten days ago.' The newspaper pursued a wildly ambitious stunt to have them adopted by the President of the United States.[46] Described as 'a diplomatic feeler', it did not stand a whisker of a chance.

Sally, the London Zoo tabby, meanwhile had had her two litters in the spring and summer, for which staff had just about managed to find homes. Then, remarkably, she had gone back into the reptile house (perhaps it was warm) and had a third litter of seven. They were found by Keeper Poole, who described them as 'a colourful batch'.

Sally herself was 'a friendly creature but the kittens bit, hissed and scratched', it was reported. Zoo staff were now 'looking for homes for them' and they were on offer to the visiting public, 'as we cannot have too many cats in the gardens, they would be a menace if they got into the aviaries'.

The seven kittens were reduced to two (it was not spelled out exactly how). One day Sally strolled in and 'was mooching about as usual' – but without her two tabby kittens. The next day Sally herself disappeared. It was assumed that she had led them off into the Park whence she herself had come back in the early spring.

46 The loopy feline plan was advanced, so *Life* magazine reported later, by the *Mirror*'s 'beauteous' New York correspondent, Georgina Campbell, along with Madison Avenue adman Robert Kendell, president of the American Feline Society. The plan was frustrated by the fact of the existence of 'Fala', the Presidential Scottie, and Mrs Roosevelt's reported dislike of cats.

Two weeks later she reappeared, 'bedraggled', with 'two kittens tumbling behind her and settled down by some hot water pipes in the boiler room'. Presumably the cold had driven her back. But her dramas were not over: Sally and her young family were in the basement of the Alligator House!

'She is too fond of romping about the reptiles,' said the keeper. 'We had a cat before who was prone to getting near the alligators to get bits of meat. But one day a nine-foot alligator snapped and almost got her.' The alligators missed out again this time. Sally and her kittens were rescued from reptilian peril. One can only hope they found safer accommodation in nearby fashionable Primrose Hill.

Chapter 28
Finest Hour

Pets recruited for the services were about to write a new chapter in military history. The War Dog School-trained patrol dogs sent to North Africa and Italy had had a tough time of it. At Northaw meanwhile, experiments had been in hand since early 1944 preparing dogs for action in the invasion of the Continent. A canine gas mask was developed, which in tests[47] dogs found to be 'perfectly comfortable', but rendered them useless in action because they could not smell anything.

That special doggy sense was paramount in a parallel programme to train former pets to sniff out land mines[48] working with Royal Engineer clearance teams. In choosing the right dog, it was noted:

> Dogs with black eyes are surly and erratic, dogs with light eyes are generally wilful, dogs with hazel

47 The dogs were held at 'Animal Farm', the Chemical Defence Experimental Station, Porton Down, Wilts, which in the course of the war consumed 15–20,000 dogs, cats, monkeys, goats, guinea pigs, etc. They too had no choice.

48 In training, a scrap of meat was concealed under the mine. Gradually the meat lure was diminished, and the dog rewarded for pointing out the now meatless mine by being given a scrap from its handler.

> coloured eyes usually have the firmer character and should be selected.

The clever Mr Lloyd was in charge of training. The most difficult to deal with were so-called 'Schu' mines, anti-personnel devices packed in a wooden box.

Mine dogs, like all good war dogs, had a trusted handler who would mark with a flag the spot of a Schu mine, on which the dog was trained to 'sit'. Results were mixed. In extreme weather dogs 'merely pretended to work'. They disliked noise and the smell of dead bodies. A report from Normandy complained about 'bitches in season being distracted by the vagrant French curs that abound in this area'. 'Handlers are far too kind,' noted a Royal Engineer lieutenant, 'nevertheless the dogs are doing excellent work.' However the unfortunate commander of a dog platoon had his foot blown off by a Schu mine that a dog had missed.

The mine hunter dogs had all once been dog's home strays or beloved pets like Michael, the Golden Retriever who had been sent away in December 1943 in answer to the war dog appeal. At their chance railway station meeting, Dr Burnell had been told their former pet was being sent to France to search for mines. Had he survived? Would he ever come home? And would he recognize the family that had loved him?

Khan, the Alsatian from Tolworth, was now 'Rifleman Khan', having passed out of Northaw and been posted to the 6th Battalion The Cameronians, the famous Scottish infantry regiment, as a patrol dog. On the night of 2 November 1944 he and his handler, Lance-Cpl Jimmy Muldoon, were fighting their way onto the Dutch island of Walcheren, a V2 rocket launch site, when shellfire hit their assault craft. Dog and handler were pitched into the icy dark waters. After a desperate search, the fearless Khan grabbed his struggling master by the collar and paddled to the mud-flat shore. 'Man and dog collapsed on the bank.' It was not the end of Khan's adventures.

Meanwhile the end of the war was in sight. The Third Reich, source of so much inconvenience for pets, was crumbling. Before too long, Adolf Hitler would retire into the Berlin bunker with Blondi, his pet German Shepherd.

Pet lovers in the services were now with the Allied armies on the German frontier. The concern of military vets was how to deal with the enormous number of enemy animals, horses especially, being taken 'prisoner'. Pets meanwhile were thin on the ground.

Private John Keet, a peacetime member of the Edinburgh General Terrier Club, was in an ENSA concert party entertaining troops. The retreating enemy had taken all the best animals, although some secretly hidden had survived, so he told *Our Dogs* of his canine observations in the winter of 1944.

The authorities in liberated Belgium were allowing 'some meat, oatmeal and bread for dogs, cats and rabbits' he reported. There were well-groomed Poodles on the streets of Brussels. Dogs were non-existent in famine-gripped Holland apart from occasional toys – Papillons, Pekes and Griffons.

Mrs Sugden of St Brelades, Jersey, would tell *Tail-Wagger Magazine* that the Germans were quite nice to the dogs on the island, giving them scraps of food, up to the end of 1944 when suddenly they all disappeared. She supposed they had been eaten.

Private Gunnar de Wit was in a cellar on the banks of the River Maas that winter, 'a far cry from the Tooting dispensary of the PDSA' where he had been a volunteer worker. 'Animals over here have had a raw deal,' he told *PDSA News*, 'Horses and cattle have been killed by shellfire, caught while they were still chained up, sometimes in wrecked sheds or in open fields.' He had to despatch a few pigs, which had been wounded by shrapnel.

'There are crazy dogs in the villages ("slap happy" the troops call them) dashing around.' He made sure they were 'disposed of humanely'.

'I have had to attend to several dogs with broken legs,' he reported. 'One of the chaps has one of them with him now, the pet of the battery.' You can guarantee he was.

Rockets fired from northern Holland continued to torment southern England. Through the autumn and winter of 1944–45, British and American bomber fleets flew almost unopposed to wreck German cities from end to end. Japan's cities burned. Pets were annihilated. Rescuers on all sides did what they could.

In London, the canine contribution to search and rescue work would be different this time. The amateur Terriers of the Blitz were yesterday's dogs; for German Shepherds, this would be their Finest Hour.

There had been abundant evidence from the Blitz that dogs (and indeed cats) were natural rescuers. If they could smell out a Schu mine, they could find a buried body. Yet in spite of their widespread training as guard and 'tactical' animals, no official rescue dogs had been used in the flying-bomb summer.

There is a story that Lt-Col Baldwin had seen a dramatic documentary in a Cheltenham cinema – *Heroic Stalingrad* – about the battle and this had given him the idea that dogs could be trained to 'point' snipers. It was certainly more complicated than that, but the idea had developed into finding buried casualties. With London now under sustained rocket attack, some doggy people thought that they might still be useful, as indeed they were.

Two 'lend-leash' pet Ministry of Air Production (MAP) guard dogs were recalled to Gloucestershire for special training in late summer 1944. Among them was Jet, the Alsatian brought up on Government rations since he was a Liverpool pup, who had lately been guarding an American airbase in Ulster. After a month of rockets killing Londoners, there was a demonstration in Birmingham in early October, where RAF volunteers hid themselves deep under rubble from 1940-era ruins. Home Secretary Herbert Morrison watched it all with senior Civil Defence officials.

At the command 'Find!' all but one was found in less than three minutes. Jet led the way. After that, it would be said, 'when Jet was satisfied that he had a find, then he would indicate it by starting to dig. He was never known to give a wrong indication, but frequently burnt his feet by the attempted digging.'

Unlike the Blitz, rockets fell arbitrarily on the giant dartboard of the capital. The outer suburbs, as well as the centre, were taking a pounding. Jet and his RAF handler, Cpl Wardle, arrived on 16 October 1944 to be based at Civil Defence Depot 1, Cranmer Court – a mansion block in Whiteheads Grove, Chelsea. Cpl Wardle's first act was to take Jet to the mortuary at St Stephen's Hospital – 'He sniffed, stepped over the bodies and took no further notice.' This was good. After all, he was supposed to be interested in living survivors. The next night in Norwood, south London, he found two under the rubble.

On the 19th 'Thorn' and MAP trainer, Mr M. Russell, and 'Storm'[49] with his trainer, Mrs Margaret B. Griffin of the famous Crumstone Kennels, arrived in London, driven

49 'Crumstone Storm' had already found pre-war fame starring in the film *Owd Bob* as 'Black Wull', the evil sheepdog champ the newcomer Collie hero has to beat (*see also* p.37).

urgently from Gloucester by Colonel Baldwin. Canine history was in the making.

From the correspondence files it is clear that veteran Civil Defence rescuers were deeply sceptical of these doggy amateurs turning up at 'incidents' in their little utility van. There was 'strong language' amid the dust, rubble and tangled limbs as the search went on for the living and the dead.

Mr Russell complained of 'very poor cooperation' at an incident at Hazlehurst Road, Wandsworth, on 19 November when the bodies of two adult males, a child and 'particles of flesh' were found (fragmentary remains were a common occurrence, something the dogs found difficult to deal with). But he predicted, 'It is only a matter of time before there is complete harmony at every incident.'

He was right. The suspicion of the old hands turned to a sense of near wonderment at what the dogs could do. Four more dogs had arrived by early November and been posted to depots in Hendon and Lewisham, expected to learn on the job. It would take a month. Mr Russell and Mrs Griffin were enrolled as part-timers in the Civil Defence and given uniforms. Margaret Griffin set up an extemporized kennel in Station Road, Loughton, in suburban Essex to cover northeast London with six dogs, including two rather special ones, 'Crumstone Irma' and 'Crumstone Psyche'. They would become very famous dogs.

Reports of the dogs in action sent into the London Civil Defence Region headquarters in the bombproof basement of the Geological Museum in Exhibition Road make harrowing reading. Bodies in a workers' canteen at Erith, Kent hit by a rocket were so fragmented it was impossible to number the casualties. In Epping, Essex, the dog discovered 'blood marks in a garden hut and later small portions of a human body. This accounts for the missing child,' as the incident report put it.

By 14 December, Irma and Psyche were ready for work. Working together in Southgate, northeast London, they found a large mound of rubble in a ruined row of houses, where Irma became excited. 'A call for silence was sufficient for a rescue officer to hear a cry from a woman saying she and her sister were in a Morrison shelter,' the incident report recounted. They were dug out, although the sister was dead. Rescue dog 'Peter' (of whom more will be heard) found three further bodies in the rubble.

The story of Miss Hilda Harvey, a schoolmistress, appeared in a newspaper the next day concerning a 'V-Bomb incident in southern England.' In very dramatic terms she told how she had been buried in rubble but could hear the 'jumbled cries of rescuers in the darkness'. Then she heard a dog sniff. 'A woman's voice said, "Leave him alone, he'll find somebody alive down there".' She heard 'the sniff again and then the bark, then the rescuers broke through the debris' and lifted her out.

No one needed further convincing. More dogs were set to training in the rubble. At the Hendon depot, the Misses M. and D. Homan were in charge of 'Rex' and 'Duke'. Mrs Griffin, now with eight dogs at the Loughton kennels, required 60 lb of meat a week to feed them and proposed getting it from the local butcher. She was informed instead she must obtain condemned meat from the Caledonian cattle market in Islington. The Waste of Food Order still applied. Rescue dogs or not, they were civilians.

Photographs show Margaret Griffin at rescue sites with her dogs, wearing a swaddling blue-serge Civil Defence greatcoat, leather gauntlets and a floppy beret with an Alsatian's head badge and shoulderflash. She is utterly magnificent, enough certainly to convince the Germans firing their beastly rockets at London that their space-age endeavour was doomed.

Mrs Griffin's own accounts are full of admiration for her animals. Irma 'had a special bark when she located someone she sensed to be alive (her "living indication").' In one incident she refused to leave a scene for two days until two young girls were found alive in the rubble. Psyche could also find pets as well as people, including 'a nice Red Setter,' badly injured but alive in a ruined house.

Mrs Griffin recalled, 'I feared his heart would give out and he could not stand up by himself. We gave him some hot tea and wrapped him in a blanket, left him quiet and telephoned for the PDSA van.'

Psyche and Irma would regularly work together. At one incident, the dogs dug around two-and-a-half feet down and found 'a lovely cat'. Irma kept giving her 'live' signal. Digging further down, rescuers 'found an old lady, her daughter and her sailor fiancé, all dead'.

'When anyone who was found by Irma was brought out dead,' wrote Mrs Griffin, 'she would try and lick the lifeless face or a hand and look up as if entreating that something more should be done.'

Mr Keith Raven was a baby when a V2 fell on Chingford in suburban Essex. With Irma and Psyche, Mrs Griffin was soon on the scene. Five years later, Raven met the heroic dogs and their trainer at Crufts and made a visit to the Goring Kennels, where he was told the story of how he survived the drama of 5 February 1945.

The row of houses in which his family had lived had been pulverized. Irma made a 'live indication' in the rubble and, as he recalled much later, 'After about 20 minutes' work, silence was called for, they could hear a baby (me) crying, after more digging and tunnelling, a way was made in and voice contact was made with our mother.' The family were buried in the Morrison shelter. Keith and his brother were found wrapped in an eiderdown and both brought out unhurt. Their mother had lost consciousness and now lay dead in the rubble.

'Peter' was a rumbustious four-year-old Collie-cross, who had been offered to the MAP for training as a guard dog by his Birmingham owner, Mrs Audrey Stable, in 1943. He was a house pet who joined up (and would go back to being a pet again) but not before he enjoyed some extraordinary adventures in London.

In spring 1945 he was the charge of the 'Rescue Man' Archie Knight based at the Chelsea depot (Peter was known as 'Rescue Dog No. 2664/9288 Peter'). The Imperial War Museum archives have the record sheet of Peter's big day, 27 March, when early in the morning one of the last rockets to be fired from Holland fell on a block of flats in Vallance Road, Stepney, inflicting severe casualties – 134 dead, 49 seriously injured. It reads:

> Report of the working dogs, Peter and Rascal [who was there for training]. This was a very big incident and the indications given were many and various. During this long period of duty Peter worked hard and well and never once refused to do all I asked of him. At the end of this day Peter was very tired.

Dog and handler were recalled to the scene the next morning. 'Peter worked very hard for two hours but was obviously affected by his exertions,' wrote Rescue Man Knight on 29 March. A second dog, 'Tailor', was not very interested – 'This was probably due to the unavoidable lack of proper food for the previous two days. Peter was completely played out and took 24 hours to recover his spirits. He is only a small dog and I consider his efforts very praiseworthy.' Mrs Maria Dickin was struck by the fact that at the rear of the smashed flats was the original oil-lamp-lit Stepney cellar where the PDSA began its work in November 1917.

On the evening of the same day the last V2 rocket fell near Orpington, Kent. The capital's pets' long ordeal was over.

*

It was getting snappy in the Berlin bunker. Hitler was allowing Blondi to sleep in his subterranean bedroom, something resented by his mistress, Eva Braun, who preferred her two Scottish Terriers, 'Negus' and 'Stasi', who were also cooped up underground. Hitler thought they were 'ridiculous' and looked 'like cleaning brushes'. He had brought them to cheer her up after a suicide attempt.

Fräulein Braun had been known to kick Blondi under the dining table. On or around 4 April, Blondi had a litter of five puppies having mated (with some difficulty) with the Alsatian, 'Harras', belonging to the architect Gerdy Troost. Hitler named the first of the male puppies 'Wulf'.

On 13 April President Roosevelt died in Warm Springs, Georgia. His Scottie, Fala, attended the funeral in Washington DC two days afterwards but, it was reported, 'seemed lost without his master'.

Five days later Berlin was bombarded by Soviet artillery for the first time. It was 20 April, Hitler's birthday. The end was near but new pets at home kept spirits high. As it would be reported later, 'Hitler was very attached to Blondi's puppies and personally fed them several times a day. The dog and the little ones had the run of the bathroom of the bunker and Hitler spent much time with them. He often took one of the puppies and then sat on the bench in the waiting room silently holding it without paying any attention to his surroundings.'

'Even Hitler set his affections on a dog which he treated very differently from the unfortunate prisoners in the concentration camps,' noted *Tail-Wagger Magazine*. How true!

In the shattered city some of the worst devastation was around the Zoo. The end was utterly tragic. One solitary hippo swam round and round a blood-red pool. The famous Pongo lay dead on the cement floor of his cage. 'Siam', the only elephant to survive the destruction of November 1943, had been driven wild by explosions and was trumpeting in terror. Meanwhile, 'Frightened apes clambered among the ruins and a few exotic birds flew from tree to tree trying to escape the acrid clouds of black smoke'.

Most of Berlin's working animals, chickens and 'balcony pigs' (rabbits) had long since been eaten. The days were numbered for those that had not been. A Berlin woman in an anonymous memoir recorded the last days:

> Someone came into the cellar with the glad tidings that a horse had collapsed outside. In no time the whole cellar tribe was in the street, which was still under [Russian] shellfire. The animal was still twitching when the first bread knife went into it.

On the afternoon of 29 April, Adolf Hitler sought to test the cyanide capsule he had been provided with by the SS. He reportedly ordered his physician, Dr Werner Haase, to give one to Blondi, aided by the Führer dog handler, Feldwebel Fritz Tornow. The dog munched it eagerly, whimpered, rolled over and died.

After midnight Hitler married Eva Braun in the Führerbunker. The next day the newlyweds committed suicide. The dog handler rounded up Blondi's puppies, scampering hopefully as they sniffed fresh air, and shot them in the garden of the Reich Chancellery. He also killed Eva Braun's Scotties (another source has one of them surviving in Munich), the dogs of the Führer's secretary, Frau Gerda Christian, and his own Dachshund.

The charred remains of Hitler, Eva Braun and two dogs (thought to be Blondi and little Wulf) were discovered three days later in a shell crater in the devastated Reichs-chancellory garden. Blondi's body was photographed by the Russians like road kill.

It was five-and-a-half years since fear of what her master had in store for them had caused Londoners to destroy almost a million of their pets. Blondi had not even been born. Millions more of Europe's innocent animals had perished in the meantime.

Epilogue

Dogs have dug into wrecked homes looking for their owners.

Cats have mewed for days outside piles of rubble, telling rescuers their owners are buried there.

Animals have quietened frightened children.

Yet when the history of the war is written, these things will not be recorded.

The Dogs Bulletin, autumn 1944

Pets Come Home

The war in Europe was over. The London Zoo's Occurrences Book was inscribed: 'Tuesday May 8th 1945, VE Day'. Arrivals that day were a Nyasaland lovebird and a blossom-headed parakeet. Departures included a 'West Indian Agouti, escaped from cage (found dead)'.

Viktor Klemperer had escaped deportation and death in the Dresden firestorm. In May 1945 he was in American-occupied Munich. Surveying the devastation, he mourned the death of his cat three years before:

> Today is the anniversary of Muschel's death, every hair of his thick little fur has been paid for by a German life.

The MAP rescue dogs Storm plus Peter and his handler were on parade at the Civil Defence stand-down ceremonies in Hyde Park on 10 June. After inspection by the King, the contingents marched past. It was reported:

> They were led by two rescue dogs walking beside their trainers. At the word of command the animals barked as they passed the royal dais. One, an Alsatian bitch, is

> credited with having located 21 living people buried in air raid incidents, and wears a medal, the other, a brown and white collie dog, saved five persons in the same way.

Clever dogs! Peter, when presented to King George VI and Queen Elizabeth, became especially excited at Her Majesty's fox-fur stole.

In Colchester, Essex, a Mrs Winifred Airlie, a pet shop owner, was prosecuted for breaking the Waste of Food Order, seemingly the last case of its kind. She had been reported for feeding bread to non-humans – a number of tame mice, the keeping of which the prosecuting counsel found 'impossible to understand'. Feeding mice was 'morally offensive' Mr Proudfoot sternly pronounced. Mrs Airlie claimed 'well-wishers posted bread through her letterbox so that she might feed her menagerie.' She was found guilty.

The end of war in the Far East and mass demobilization meant many happy reunions for returning servicemen with pets they had left behind. But Whitehall officialdom quailed at the prospect of a global zoo of adopted overseas pets also heading for British shores.

From far-off Burma came the report: 'Never has there been such a diversity of livestock as is to be found to-day in the Fourteenth Army.' Along with bears, sheep, snakes parakeets, monkeys and Siamese cats, 'Indian Army officers have brought Dachshunds, Bull Terriers, and Spaniels with them. Lieutenant-General Sun Li-jen, commanding the Chinese First Army, has an Alsatian called Mogaung and no fewer than six puppies.

'Small monkeys are to be seen riding with the drivers of trucks. Nearly every unit has a little poultry farm of its own.' One such 'farm', full of ducks and turkeys, had been created in the back of a 15 cwt truck, whose driver was 'anxious to take them back with him to Derbyshire when his term of service is up'.

But there was a way for servicemen to get pets home. A 'Special Services Scheme' for the quarantine of pets began to be discussed between the War Office, Admiralty and Air Ministry, in autumn 1944, by which 'a limited number of dogs' might be imported under special licence from the Army veterinary authorities. Quarantine facilities were to be created at a former airfield at Chilbolton Down, near Andover, with subsidized fees depending on rank (£20 for senior officers, £5 for other ranks) – 'Commanding officers must be satisfied that the dog in question is a genuine pet.' Royal Army Veterinary Corps clearing stations for service pets were established at Antwerp and Milan. According to the history of the Corps: 'The disposal of surplus horses and mules involved much work,[50] but nothing gave more trouble than the repatriation of dogs, both the pets of returning soldiers and the hundreds of dogs loaned by the public for war service.' Not everyone was happy: dog columnist Huldine Violet Beamish expressed her 'horror' in *Country Life* that servicemen should be allowed to bring 'the various canine oddments collected in Europe and the Near East back to this country', to add to the already over-large mongrel population.

The Government scheme was terribly benign really, considering. But it was dogs only. Cats ('and all other feline animals'), returning pet lovers would have to pay

50 The British Army had 120,000 animals on charge in 1945, plus 80,000 captured 'enemy' horses and mules. Within a year this was reduced to 27,000 by a policy of approved sales and humane slaughtering. The scandal of selling old warhorses locally, as in 1918, was not to be repeated.

for themselves. Swine, ungulates and anything more exotic were to stay out altogether.

The Royal Navy was presumably going to have to ship them all home. A sympathetic admiral noted, 'I consider we should not forbid the use of HM ships, assuming that the dogs it is desired to land in this country may have been pets on board for some months.' They would however be limited to four per ship, confined to kennels on the upper deck, and, 'be the responsibility of the ship's butcher'. An excellent arrangement!

In April, the RSPCA, not to be upstaged by the PDSA with their Dickin Medal, asked Brigadier George Kelly, the Army's chief vet, to supply details of the war careers of pets loaned by the public for the duration especially deserving of commendation so that they could be returned to their owners with 'a collar and medallion'. The War Office said it had no such information but would endeavour to find out. It would be a little while in coming.

And so the commendations came in. Brian from Loughborough and Bing from Rochester, the two Alsatians that had parachuted into Ranville, Normandy on D-Day, were recommended 'For Valour'. 'Ricky', a Welsh Collie, had been wounded in the head while mine hunting by the Nederweert Canal in Holland but had carried on working.

'Scamp' of Rock Ferry, Cheshire found twenty-five Schu mines in the same operation. 'Raf', an Alsatian from Erdington, Birmingham, had been 'blown up during the Ardennes push'. He was now recovered and working with a Corps of Military Police (CMP) Company. Brave dogs all!

Other dogs had been engaged guarding HM Government stores. There was 'Blackie', an Alsatian from Maryhill, Glasgow, who had in July 1944 beaten off eight thieves attempting to burgle a clothing store. Blackie had rendered invaluable assistance to his handler in pinioning the chief malefactor and 'taking random bites at the other Italian civilians'.

'Piggy', a Boxer employed by the military police in Cairo, had attacked four 'native thieves' and lost an eye in the fray but had carried on devotedly. She had now recovered and returned to duty.

'Snook', a Doberman Pinscher, had been blown up by a land mine in the Egyptian desert in pursuit of more native thieves but was now back on duty.

But how would these combat veterans take to civilian life? Would they remember their families, would their families remember them? The War Dog Training School had transported itself from Potters Bar, north London, to the Continent in late 1944 and was now set up in a former German Army tank range at Sennelager with a full complement of ATS kennel maids. Numbers of ex-Wehrmacht Schäferhunde swelled the ranks, along with sundry Belgian Bouviers de Flandres and Malinois liberated along the way. There were several marriages celebrated at the garrison church, with war dogs providing the guard of honour. It was noted: 'Now that hostilities have ceased the War Office has no authority to hold these dogs. They have no further training but are handled and exercised to make them more amenable as pets on return to their owners.'

But would they be? 'Brutus', a demobilised Alsatian from Syston, Leicestershire, met his master at the railway station after three years of war service, but 'failed to recognise him', according to *The Dogs Bulletin*. But when he reached home 'he recognised the voice of his mistress at once and was in transports of joy. With the two children aged five and two, he at once became a firm friend and playmate.'

Michael, the Burnell family's Golden Retriever, had been out of their lives for three years as a mine dog. Other than his handler, troops had been under strict orders not to make him a pet. A letter arrived asking if they would like him back. One day he appeared at the door with his handler. As Elizabeth Burnell told the story:

> Michael, on hearing my mother's voice, broke free from his lead, shot through the gate and flew into the house and up the stairs to where my mother was standing. He jumped up, put his paws on her shoulders and licked her face as she started to cry with joy. He then came and sniffed all round us.

Of the twelve mine dogs who had gone to France, only eight came back. Michael was one of them.

The tales of heroic animals tumbled out, pounced on by an eager press. Jet, Beauty (the PDSA dog) and Irma all received the Dickin Medal on 12 January 1945, while Thorn was presented with his at the Civil Defence football match at Wembley on 2 March.

The civilian animal memorial that Maria Dickin had proposed was quietly forgotten. Who wanted to remember all that upsetting stuff? The PDSA's north London organiser, former concert singer Dorothea St Hill Bourne, published a book about hero war animals in 1947 and it was a huge hit. The same year, Jet and Judy, Dickin medalists, made fundraising appearances with the film actresses Jane Walsh and Norah Swinburne, 'at various London West End stores and restaurants'. Edward Bridges Webb's populist agenda triumphed in the glow of victory. And why not? At least animals got a mention.

Little Rip, the Poplar star of 1940, got a medal, along with Rex and Peter, the MAP Alsatians of the V2 episode.

War dog 'Rob', another parachuting dog, was awarded for bravery under fire and 'Sheila', a Border Collie who rescued four US bomber crewmen from a crash in a snowstorm in December 1944 got the Dickin Medal.

Brian the parachuting dog had gone into action again in March 1945 during the Rhine crossing. Described by an RAVC major as an 'obedient dog who responded to the words of command at once', he remained on military duty until April 1946. Brian returned to Britain, where he spent his time at the Chilbolton quarantine kennels – where it was noted he had a broken tooth and scarred leg. He went home in October 1946 and seemed to have adjusted readily enough to a less eventful life. On 26 April 1947, he was awarded the Dickin Medal in a special ceremony held in London.

Rifleman Khan, the hero of Walcheren, was also nominated by the battalion commander for the Dickin Medal, presented on 27 March 1945 at a full battalion parade. The citation read: 'For rescuing Corporal Muldoon from drowning under heavy shellfire during the assault at Walcheren, November 1944, while serving with the 6th Cameronians.'

Corporal Muldoon wrote to the War Office asking to be allowed to keep Khan after the war ended. The Railton family also asked for their dog to be returned. The Alsatian was now that rare thing: a tug-of-love war pet.

Corporal Muldoon returned to civilian life in Scotland. Khan was in quarantine not far from his original home. Barry Railton, now twelve, visited Khan three times a week. At the end of six months, veteran Khan was returned to leafy Tolworth.

In August 1947, the plucky Alsatian was invited to participate in a War Dog Parade at Wembley Stadium. Harry Railton wrote to Muldoon via the War Office, asking him to lead Khan in the parade. The Scot was overwhelmed at the thought of seeing Khan again – 'Two hundred of the most intelligent, skilful dogs in Britain, including sixteen Dickin medallists, were to appear.' In a moving account of the big day:

> Khan was one of a huge crowd of dogs milling around. Suddenly he stopped, lifted his head, his ears at the alert. He sniffed the air. His legs tensed. He jerked the leash from Mr. Railton's hand and bolted, a streak of fur, across the parade ground, barking loudly.
>
> Ten thousand people in the viewing stand saw the joyful reunion of man and dog. Applause thundered as Muldoon and Khan took their places in the parade line. Afterwards Harry Railton searched out Muldoon in the crowd. He watched as Muldoon, tears bathing his cheeks, buried his head in the dog's fur. Sobbing, he held out the leash. Mr. Railton shook his head. 'Barry and I talked it over during the parade,' he said. 'Tell him, Barry.' 'We think Khan belongs with you,' said Barry. 'He is yours. Take him home.'

Faith, the church cat who tended her kittens in the 1940 Blitz, 'could not be given a Dickin medal because she was a "civilian" and not with the armed forces or civil defence'. But, as it was said shortly afterwards, 'Mrs. Dickin had been so impressed with Faith's bravery that she said she was going to have a special silver medal made and present it herself.' The Archbishop of Canterbury (not the one who three years previously had vetoed the wartime prayer for animals) was at the 12 October 1945 ceremony. It was recorded:

> While the Archbishop held the cat, Mrs. Dickin read the citation from a parchment scroll, and presented the

medal, hanging it around Faith's neck. The Archbishop gave a brief address, and then everyone retired to the adjoining hall for a buffet lunch. The parishioners had saved their rations and all contributed to a splendid spread. Faith had a special plate of fish.

Faith's fish in fact broke the Waste of Food Order, which was still in force. The question of what to feed them would still test the ingenuity of pet owners. Offal and sausages, restricted since 1943, had come off ration but meat would remain state-restricted for almost a decade to come. Bread and potato rationing were yet to begin; dog biscuits were thin on the ground. Pets were getting impatient.

Fox hunting went through another austerity winter. In fact a MAFF official would admit in June 1945 that, when three years before rations for hounds and horses had been reduced, the 'reason was largely political'. It made no difference to the real animal feedstuff resources. 'The supply position is still very difficult', he noted; however, 'anything which might create an impression otherwise is to be avoided.'

Horse & Hound complained of seeing a 'fine, fat fox waddling' out of a storm drain and there being nothing the hunt could do about it. The fox's fatness seemed a special irritant. The December 1944 article looked forward to a comeback of fox coverts that had been ploughed up or 'absorbed into aerodromes'.

The first post-war edition of *Baily's Hunting Directory* mourned the casualties of war: 'The Earl of Harrington, the Afonwy and Aberdeenshire harriers have been left without a hound. The entire pack of the Buckinghamshire Otterhounds has been dissolved. The Barrington, Brissenden and Wideford Beagles have ceased to exist.'

It noted gratefully how farmers had shown themselves 'determined that hunting should continue with the gift of carcasses unfit for human consumption' to feed packs of hounds.

The post-war fox meanwhile showed his cunning in making 'the most skilful use of dumps of rusting barbed wire and enclosed aerodromes'. The setbacks of the war were nothing compared with the much more serious threat – 'the increasing momentum of anti blood-sport propaganda which is now a private member's Bill'. It would be defeated.

The animal charities took the coming of peace as the chance to return to their glum civil war. The RSPCA and the Canine Defence League refused to attend a grand conference in October 1945 called to address animal welfare politics.

Old habits were hard to break. As Parliament debated the issue, the RSPCA could not condemn fox hunting. Our Dumb Friends' League tried to revive the plight of old British war horses but was told practically no animal had been left alive by the retreating Germans. Any that remained had probably been eaten. In the British Control Zone of Germany moves were taken to ship surviving German Zoo stock to Britain so as to leave more food for humans. The de-Nazification specialists in the Military Government wound up the Reichtierschutzbund but left the Reich animal protection laws of 1933 in place. Army vets with the British control commission found that the German civil veterinary service, headed by Dr Friedrich Weber, a close friend of Hitler, 'was one of the most pro-Nazi branches of civil administration in Germany'.

The Animal Guards disbanded without ceremony. The ARP Animals Committee mutated into something called the 'National Registration of Animals Service' before disappearing altogether, leaving some identity disks in cupboard drawers as curious reminders of pets at war.

The Technical Officers of the PDSA, so long derided as unqualified quacks, received qualified recognition by parliamentary legislation.

Europe's zoos were in ruins. A visit by *The Times* to Berlin in early 1946 discovered, 'the animals, few and far between, must be tracked down amid ruined buildings and piles of rubble':

> The elephant looks a rather woebegone creature. There are few buns or apples to spare for him just yet. Other survivors from the days of bomb and battle include two young lions, and there is as well a fine male lion, a newcomer since the capitulation. A lone chimpanzee delights young Berliners by riding in a swing and other antics. Two animals had to be killed to make food for the others, but none has starved.

Frau Doctor Heinroth, widow of the former head of the Aquarium, was in charge. The Zoo appealed to patrons to help provide food and was offering the following prices: '10 pfennigs for a mouse, 20 pfennigs for a rat, 8 pfennigs for a kilo of acorns and 1 mark 50 pfennigs for a litre of worms.'

A 1946 statement from the once-proud Hagenbeck Tierpark in Hamburg described the daily requirement for seals, pelicans, etc. as 80 kg of fish a day. For the carnivores it was 100 kg. But it was not getting through. Herbivorous animals subsisted on turnips and ruminants on 'acorns collected by Hamburg school children'. The hippo was starving. It was decided to ship it to London, along with two more and an elephant from Hanover Zoo.

In London the physical damage to the Zoo seemed

'negligible' (although the Zebra House still lay in ruins), while numbers of visitors were hugely up over 1939. The end-of-1945 annual report stated that there were 1,849 animals at Regent's Park, and 824 at Whipsnade, their numbers boosted by all those donated mascots and regimental pets.

Wartime rescuers left the stage. The Duchess of Hamilton died on 12 January 1951, leaving the Ferne estate in Wiltshire as an animal sanctuary. Mrs Maria Dickin died aged 80 on 1 March 1951 and Edward Bridges Webb became chairman of the PDSA. Louise Lind-af-Hageby continued to run Ferne and Animal Defence House in St James's Place with inevitably dwindling competence. She died at 7 St Edmunds Terrace on 26 December 1963, leaving her fortune to the Animal Defence Trust. The legacy became the matter of a financial scandal and a Department of Trade & Industry investigation of a Dartmoor badger sanctuary.

Ferne House was demolished in 1965, along with the centrally heated aircraft hangar that so comforted evacuee cats in the great emergency of 1939–41. A new house would eventually be raised in neo-Palladian style for the proprietor of a cat-supporting tabloid newspaper. Ferne Animal Sanctuary, as it is still called, continues at another site near Chard in Somerset. The London pet refuge, 'Lyndhurst', and next door No. 7 St Emund's Terrace were replaced by a block of flats in the seventies. The Lido Cinema, Bolton, was demolished in 2006. Animal Defence House in Mayfair, once crammed with refugee pets, was home to a hedge fund seventy years later. Hook House, near Potters Bar, the Army War Dog School, became the UK Oshwal religious centre in 1989.

Rota, the lion from Pinner presented to Winston Churchill in 1943, sired forty cubs. He died in 1955, was stuffed and mounted then purchased for display by a Florida museum.

One by one, the war hero animals, celebrities all, left the field. The little Poplar Terrier, Rip, died in 1946 and was buried in the PDSA's Ilford cemetery. His headstone was inscribed 'Rip, D.M., "We also serve" – for the dog whose body lies here played his part in the Battle of Britain'. Beauty, the pioneer search and rescue dog, died on 17 October 1950 and was duly laid to rest. The death of Faith, the Blitz cat, in September 1948 was reported in *The Times* ('little is known of the life of Faith before she entered the church, she was a stray') where her bravery was compared with Mourka, the Stalingrad tabby. She was buried in St Paul's churchyard.

Tich, the desert adoptee, got to Ilford in 1959 ('Sleep Well Little Girl') to be buried in a flag-draped coffin with full military ceremonial to lie alongside 3,000 no less adventurous wartime pets. They included 'Hitler', 'Timoshenko', 'Joffre', 'Dinky', 'Dusty', 'Ginger' and 'Peter', the Home Office cat.

Brian, the parachuting D-Day Alsatian from Loughborough, died aged 13 in 1955 and was buried alongside his fellow Dickin Medal holders in Ilford, Essex, close to the field where so many less celebrated pets had been interred, unnamed and unmemorialized, in the great destruction panic on the outbreak of war and the mass abandonment of late summer 1940. That is where little Oo-Oo and Bonzo went – along with hundreds of thousands of cats and dogs, the ordinary pets of Londoners. It is a haunted place.

The overgrown cemetery in east London was restored in 2007 with National Lottery funding, but the charity 'chose not to emphasise' the fact or manner of their passing.

But they will be remembered.

Acknowledgements

I am grateful to the archivists, librarians and charity gatekeepers who kindly made available so much hidden information on wartime pets. Special thanks to Francesca Watson and her colleagues at Cats Protection (CPL), Tracey Hawkins of Blue Cross (ODFL), Chris Reed of the Royal Society for the Prevention of Cruelty to Animals, Michael Palmer of the Zoological Society of London, Clare Boulton, librarian of the RCVS, Ruth Macleod of Wandsworth Heritage Service, Jean Rosen of the IWM Department of Documents, Sarah Graham of the Dogs Trust (NCDL) and Mr Alan Meyer for information on the Animal Defence Trust.

I am additionally grateful to these organisations plus the PDSA for the reproduction of material from their contemporary publications and reports. I am grateful to the publishers of *The Dog World* and *Our Dogs* for permission to quote from wartime issues of their publications. The *Manchester Guardian* newspaper is quoted by permission of the Guardian Media Group, the extract from *Boy in the Blitz* by Colin Perry by permission of his daughter, Felicity Collier, and extracts from the published works of Norman Longmate, Lionel Montague and Arthur W. Moss appear by permission of the Random House Group. Quotes from the *Diary of George Orwell* are copyright George Orwell and appear by permission of the Random House Group. The extract from the poem 'War Cat' by Dorothy L. Sayers is by permission of David Higham Associates.

Transcripts from Crown Copyright documents in the National Archives at Kew appear by permission of the Controller of HM Stationery Office. Extracts from Archbishop Temple's wartime correspondence on prayers for animals are by permission of the trustees of Lambeth Palace Archives.

I am grateful to the Trustees of the Imperial War Museum and the individual copyright holders for granting access to the collections of private papers held by the IWM and for permission to publish extracts from them. Mass Observation (M-O) material is reproduced with permission of Curtis Brown Group Ltd, London, on behalf of the Trustees of the Mass Observation Archive, Copyright © the Trustees of the Mass Observation Archive.

Copyright in the extracts from the WW2 People's War online archive of wartime memories contributed by members of the public belongs to the contributors as credited, licensed to the BBC. The archive can be found at www.bbc.co.uk/ww2peopleswar. Every effort has been made to trace the copyright holders of further unpublished documents and published works in print or electronic form from which quotations have been made.

Thank you especially to two remarkable German Shepherd devotees, Joyce Ixer and Garbo Garnham, for sharing wartime memories.

I would also like to thank my publisher, James Gurbutt, and agent Felicity Blunt for getting behind this book so enthusiastically from the start, as well as editors Jane

Donovan and Clive Hebard for their intelligent and insightful comments.

In addition, my thanks to Victoria and Joseph, who tirelessly run their own cat rescue project in Clapham, and from whom we received our own two beloved felines, Fergus and Luis, and who along with my aunt's poor, long-gone Paddy, were part of the original inspiration for this book.

Finally, I would like to thank my husband and co-writer, Christy Campbell, whose hard work, enthusiasm and eye for detail never wavered, and our daughters, Maria and Katy, and son Joseph, who have inherited their parents' passion for pets and concern for their jeopardy.

Bibliography

Ackerman, Diane, *The Zookeeper's Wife* (W. W. Norton & Company, New York, 2008).

Baily's Hunting Directory (J. A. Allen, London, 1949).

Ballard, Peter, *A Dog Is For Life* (National Canine Defence League, London, 1990).

Bondeson, Jan, *Amazing Dogs* (Amberley, Stroud, 2011).

Bourne, Dorothea Saint Hill, *They Also Serve* (On Animal Mascots In The Allied Forces) (Winchester Publications, London, 1947).

Calder, Angus, *The Myth of The Blitz* (Pimlico, London, 1992).

Chance, Michael, *Our Princesses and their Dogs* (John Murray, London, 1936).

Charman, Terry, *Outbreak, the World Goes to War* (Imperial War Museum, London, 2009).

Clabby, J., *The History of the Royal Army Veterinary Corps, 1919–1961* (Allen, 1964).

Clarke, Maurice, *First Aid to Dogs and Cats: A Handbook for the Technical Personnel of the National A.R.P.* (For Animals Committee) (Baillière, London, 1941).

Colville, John, *The Fringes of Power* (2 vols) (Sceptre, London, 1986–87).

Cooper, Jilly, *Animals in War* (Corgi, London, 1984).

Dickin, Maria Elisabeth, *The Cry of the Animal* (PDSA, London, 1950).

Douglas, Nina Mary Benita, Duchess of Hamilton and Brandon, *The Chronicles of Ferne* (The Animal Defence Society, London, 1951).

Edwards, Thomas, *Regimental Mascots and Pets* (Hutchinson, London, 1940).

Giese, Klemens, and Waldemar Kahler, *Das Deutsche Tierschutzrecht* (Berlin, 1944).

Graves, Charles, *Great Days* (Hutchinson, London, 1944).

Gray, Thelma, *The Corgi* (Chambers, 1952).

Heck, Lutz, *Tiere-Mein Abenteuer* (Animals – My Adventure) (Methuen & Co., London, 1954).

Home Office (ARP Dept.), *Air Raid Precautions Handbook No. 12* 'Air Raid Precautions for Animals' (HMSO, London, 1939).

Huxley, Julian, *Memories* (Allen & Unwin, London, 1970).

Jenkins, Garry, *A Home of Their Own* (Bantam, London, 2011).

Kean, Hilda, *Animal Rights* (Reaktion, London, 1998).

Keeling, C. H., *They All Came Into the Ark* (Shalford Clam Publications, 1988).

Kelway, Phyllis, *The Squirrel Book* (Collins, London, 1944).

Klemperer, Viktor (ed. Martin Chalmers), *The Diaries of Viktor Klemperer 1933–1945* (Phoenix, London, 2000).

Lind-af-Hageby, Louise, *Bombed Animals-Rescued Animals-Animals Saved From Destruction* (Animal Defence & Anti-Vivisection Society, London, 1941).

Longmate, Norman, *How We Lived Then* (Arrow, London, 1973).

—, *The Doodlebugs* (Hutchinson, London, 1981).

—, *Hitler's Rockets* (Hutchinson, London, 1985).

Montague, Frederick, *Let the Good Work Go On* (On the Work of the People's Dispensary for Sick Animals of the Poor) (Hutchinson, London, 1947).

Moss, Arthur W., *Animals Were There: A Record of the Work of the RSPCA* (Hutchinson, London, 1947).

Orwell, George (Ed Peter Davison), *Diaries* (Harvill Secker, London, 2009).

Reynolds, Quentin, *All About Winston Churchill* (W. H. Allen, London, 1964).

Richardson, Edwin H., *British War Dogs, Their Training and Psychology* (Skeffington, London, 1920).

RSPCA, *Animals and Air Raids* (2nd edn) (RSPCA, London, 1939).

Sauerbruch, Ferdinand, *Das War Mein Leben* (Bad Wörishofen, 1951).

Sax, Boria, *Animals in the Third Reich* (Continuum, London, 2000).

Schwangart, Prof Dr Friedrich, *Vom Recht der Katze Leipzig* (Paul Schöps, 1937).

Smith, Carmen, *The Blue Cross at War* (Blue Cross, Burford, 1990).

Stephanitz, Max von, *The German Shepherd Dog in Word and Picture* (Fairholme, Uckfield, 1925).

Stone, Christopher, *Salute to the National Animal Guard* (National Registration of Animals Service, London, 1945).

Studnitz, Hans-Georg von, *While Berlin Burns* (Weidenfeld & Nicolson, London, 1964).

Thompson, John, *The Animals Go Too* (Gawthorn, London, 1939).

Turner, E. S., *The Phoney War on the Home Front* (Michael Joseph, London, 1961).

Source Notes

Introduction

p. xiii 'All the animals that served …' (www.animalsinwar.org.uk).

A Note on the Animal Welfare Charities

p. xvi 'Anyone who knows anything …' Report of Inquiry into Affairs of NARPAC, November 1940 (TNA HO 186/2075).

p. xvii '[Our cat] shelters should' ODFL report for 1929 (p. 54) London Metropolitan Archives A/FWA/C/D256/4

p. xvii footnote '140,000' Jenkins, *A Home of Their Own* (p. 242), police figures for dogs destroyed are in TNA MEPO 2/6597 16 October 1945.

p. xviii 'anyone who devotes …' *Daily Express*, 11 May 1938 (p. 12).

p. xviii 'One East End woman …' Montague, *Let the Good Work Go On* (p. 94).

p. xxi 'We feel that people …' *The Times*, 18 June 1938 (p. 16).

p. xxii 'The P.D.S.A. Rescue …' PDSA Annual Report, 1945 (p. 25).

p. xxii 'thousands of animals …' PDSA Annual Report, 1946.

p. xxii 'too far removed …' M. R. Curtis to Maj-Gen Peter Davies 15 December 1998 IWM National Inventory of War Memorials AIW Box 4.

p. xxii 'If you could offer' Marliyn Rydström to Maj-Gen Peter Davies 2 January 2002 IWM NIWM AIW Box 12.

p. xxiii 'the Opposition …,' Sir Charles Hardinge …' 8 June 1939 (TNA MEPO 2/6478).

PART ONE: PAWS IN OUR TIME

Chapter 1: In Case of Emergency

p. 3 'In Memoriam', *Tail-Wagger Magazine*, October 1939 (p. 340).
p. 6 'We urged all these ...' Our Dumb Friends' League Report for 1938.
p. 6 'Ensure the safety ...' Kennel, Farm and Aviary, *The Times*, 3 October 1938 (p. 3).
p. 7 'Send your dogs ...' ibid., 29 September 1938 (p. 3).
p. 7 'to inform owners ...' ibid., 30 September 1938 (p. 1).
p. 7 'In the event of war ...' ibid., 29 September 1939 (p. 9).
p. 7 'Black Cat No. 10 ...' *The New York Times*, 5 November 1938 (p. 3).
p. 8 'The experience of Spain ...' *Manchester Guardian*, 21 May 1938 (p. 14).
p. 8 'to secure and shoot a cat ...' *RSPCA Animals and Air Raids* (p. 3), TNA MEPO 2/6478).
p. 8 'po-faced rightwing bureaucrat ...' Calder, *The Myth of the Blitz* (p. 47).
p. 9 'When the good news ...' *The Animal World*, November 1938 (p. 1).
p. 9 'During the 48 hours ...' *The Times*, 22 May 1939 (p. 20).
p. 9 'In the animal ...' *Luftschutz der Tiere und Verpflegungsmittel* Prof Dr C. E. Richters, 1938 (http://www.bunker-dortmund.de/index.php?page=363).
p. 10 'The last few days of September ...' *The Dogs Bulletin*, October–November 1938 (p. 6).
p. 10 'During the war-crisis of September ...' Lind-af-Hageby, *Bombed Animals*, 1941 (p. 13).
p. 11 'Mrs Freeman sent six cats ...' Douglas, *Chronicles of Ferne*, 1948 (p. 62).
p. 11 'week of throbbing tension ...' *The Dogs Bulletin*, Oct–Nov 1938 (p. 6).
p. 11 'Another inquirer ...' *The Animal World*, December 1940 (p. 100).
p. 11 'We have had numerous ...' *Cat World*, 1 October 1938 (p. 1).
p. 12 'One of our members ...' 'Notes of the Month', *The Cat*, June 1938 (p. 1).

p. 12 'Until we know for certain …' *The Cat*, October 1938 (p. 75).
p. 12 'Those of us who …' ibid. (p. 76).
p. 12 'One feature of the crisis …' *The Dogs Bulletin*, October–November 1938 (p. 6).
p. 13 'The League felt …' Smith, *Blue Cross at War* (p. 46).
p. 13 'As one penniless …' *The Dogs Bulletin*, February–March 1939 (p. 4).
p. 13 'refugee dogs …' LAPAVS News-sheet, August 1940 (p. 4).

Chapter 2: It Really is Kindest …

p. 14 'commandeered healthy horses …' RSPCA Report for 1939 (p. 46).
p. 14 'animals in a …' 27 January 1939 (MEPO 2/6478).
p. 15 'rabbits, guinea-pigs …', ibid.
p. 15 'dogs dressed in Angora …' The *Daily Mail*, 9 February 1939 (p. 8).
p. 16 'We two children …' Our Dumb Friends' League Report for 1939 (p. 21).
p. 16 'Many of these refugee dogs …' *The Dogs Bulletin*, June–July 1939 (p. 2).
p. 17 'a really lovely striped tabby …' *The Cat*, June 1941 (p. 89).
p. 17 'His morning bird walk …' *Animal and Zoo News*, December 1939 (p. 22).
p. 17 'Neville Chamberlain Moth Hunter …' *Animal and Zoo News*, April 1940 (p. 8).
p. 18 'Cattle and sheep …' 14 April 1939 (TNA HO 186/1417).
p. 18 'the compulsory slaughter …' (TNA MEPO 2/6478).
p. 18 'because in an air raid …' *Bee Craft*, April 1939 (p. 101).
p. 18 'Besides the 40,000 …' *The Times*, 22 May 1939 (p. 20).
p. 19 'The RSPCA …' April 1939 (TNA MEPO 2/6478).
p. 19 'One is well aware …' ibid.
p. 20 'people with cars …' ibid.
p. 20 'strays in 'good condition …' ibid.
p. 21 'Dogs and cats …' (TNA HO 45/18150).
p. 21 'when an owner …' *ARP Handbook No. 12* – 'Air Raid Precautions for Animals' (p. 24) (TNA HO 186/2489).
p. 21 fn 'I attended a meeting …' (TNA MEPO 2/6478).
p. 22 'an extension of …' ibid.

p. 22 'evacuation would ...' 22 June 1939 (pp. 1–6) (TNA HO 186/1418).
p. 22 'carcasses, once collected ...' ibid.
p. 22 'thought the outbreak ...' ibid.
p. 22 'cat and dog lethalling ...' (TNA MEPO 2/6478).
p. 23 'two kinds of identity ...' ibid.
p. 23 'Temple Cox ...' 17 March 1939 (TNA MEPO 2/4413).
p. 23 'primary task ...' (TNA HO 186/1418).
p. 24 fn. 'the least an owner ...' 27 June 1939 (TNA MEPO 2/6478).
p. 24 'There must be no suggestion ...' 4 August 1939 (TNA MEPO 2/6478).
p. 25 'Mr Badger ...' 1 June 1940 (TNA HO 186/1418).

Chapter 3: Killed by Order

p. 25 'Gas proof kennels for dogs ...' *Tail-Wagger Magazine* November 1939 (p. 364).
p. 25 'contaminated hair on cats ...' *ARP News* October 1939 (p. 23).
p. 25 'goldfish mildly off colour ...' ibid.
p. 26 'Hang caged birds ...' *The Times* 8 September 1939 (p. 10).
p. 26 'In the first place ...' *The Veterinary Record* 5 July 1941 (p. 390).
p. 26 'It is obviously impossible ... brochure for The Frank-Heaton Protective Enclosure against Gas, Splinters & Blast For Small Animals Sussex University M-O Archive Topic Collection 79 Dogs in War Time 1939-42.
p. 26 'As a last general hint ...' *The Veterinary Record* 5 July 1941 (p. 390).
p. 27 'I enclose two snaps ...' M-OA T C 79.
p. 27 'YOUR FOOD IN WAR-TIME ...' Public Information Leaflet No. 4 July 1939 http://squirrelintheattic.blogspot.co.uk/2006/12/your-food-in-wartime-public-information.html (p. 2).
p. 28 'I have had the joy ...' *The Dogs Bulletin* June-July 1939 (p. 2).
p. 28 'Twice a week ...' Keeling, *They All Came into the Ark* (p. 130).
p. 29 'Only prosperous classes ...' *The Animal World* October 1939 (p. 189).
p. 29 'Cats are not ...' ibid.
p. 29 'In view of the present emergency ...' (TNA MEPO 2/6478).
p. 30 'Measures to Meet an Immediate Emergency ...'(TNA HO 186/1417).

p. 30 'exactly conforming …' ibid.
p. 31 'Official Advice …' Care of Pets *The Times* 26 August 1939 (p. 15).
p. 32 'the Animal Defence …' Lind-af-Hageby op. cit. (p. 17).
p. 32 'Homes in the country …' *The Times*, 29 August 1939 (p. 1).
p. 32 'The result of the broadcast …' Douglas, op. cit. (p. 14).
p. 33 'go out in its name …' (TNA HO 186/1418).
p. 33 'Is there a pet in the house?' *Daily Mirror*, 28 August 1939 (p. 12).
p. 34 'A passer-by gave him …' *Daily Express*, 26 August 1939 (p. 7).
p. 34 'mystery black cat …' *Daily Express*, 28 August 1939 (p. 1).
p. 35 'I seem to see such …' Mrs G. H. Langford (M-O Diarist D5350).
p. 35 'In case they ran amok …' Longmate, *How We Lived Then* (p. 30).
p. 35 'Whatever happens …' *The Dog World*, 1 September 1939 (p. 1).
p. 36 'What this will mean …' *The Kennel Gazette*, 1 September 1939 (p. 1).
p. 36 'Overseer MacDonald', *Evening Standard*, 3 September 1939 (ZSL Newscuttings Vol. 36).
p. 37 'The Border Terrier Bitch', *The Dog World*, 27 December 1940 (p. 1126).
p. 37 'Smout', 2 September 1939, Brinton Lee (M-O diarist D5262).
p. 37 'the principal routes …' Mollie Panter-Downes, *The New Yorker Magazine*, 9 September 1939.
p. 38 'Germany invades Poland …' ZSL Occurrences Book, 1 September 1939.
p. 38 'When the news …' Huxley, Memoirs (p. 249).
p. 39 'The poisonous snakes …' *Daily Telegraph*, 2 September 1939.
p. 39 'her dog got out of its basket' Longmate op. cit. (p. 26).
p. 39 'Gardens closed at 11.00 …' ZSL Occurrences Book, 3 September 1939.
p. 40 'practically valueless …' Keeling, op. cit. (p. 134).
p. 40 'bored and disgusted …' *The Sunday Times*, 10 September 1939.
p. 40 'Koringa' ZSL Occurrences Book, 9 November 1939.
p. 41 'The people left in London …' *Animal and Zoo News*, February 1941 (p. 13).
p. 41 'had paid sixpence …' *The Animals' Defender*, June 1940 (p. 13).
p. 41 'It is not as the proprietor …' *Animal and Zoo News*, February 1941 (p. 13).

Chapter 4: Killed by Kindness

p. 42 'large number of household ...' Anon t/s in RSPCA archives.
p. 43 'On Sunday 3 ...' ibid.
p. 44 'the best thing for animals ...' *The Ilford Recorder* 7 September 1939 (p. 4).
p. 44 'PDSA performed ...' PDSA press office statement 22 April 2013.
p. 45 'over two and a half million ...' 11 September 1939 (TNA HO 186/1418).
p. 45 'The committee urge owners ...' ibid.
p. 45 'When the embassy closed ...' *Daily Express*, 6 September 1939 (p. 7).
p. 45 'ham sandwich ...' ibid.
p. 45 'now recovering ...' 'How Now Brown Chow', *Daily Express*, 8 September 1939 (p. 6).
p. 46 fn 'I can hardly conceive ...' Sir Nevile Henderson, *Hippy*, 1943.
p. 46 fn 'which bore out stories ...' Montague op. cit. (p. 88).
p. 46 'By the dog that Ribbentrop ...' *Daily Mirror*, 7 September 1939 (p. 10).
p. 46 'insulting remarks ...' *The Animal World*, October 1939 (p. 187).
p. 46 'A widespread and persistent ...' *The Times*, 7 September 1939 (p. 3).
p. 46 'huge destruction of cats ...' ibid.
p. 46 'All estimates were ...' (TNA HO 186/1418).
p. 47 '80,000 in one night ...' *Sunday Express*, 7 April 1940 (p. 11).
p. 47 'With the advent of war' (TNA HO 186/1418).
p. 47 '¾ of a million' ibid.
p. 47 'The stories I am hearing ...' *The Dog World*, 15 September 1939 (p. 570).
p. 47 fn 'It was the greatest single ...' *Sunday Express*, 7 April 1940 (p. 11).
p. 48 'Enemy Alien', *Woman*, 18 November 1939 (p. 12).
p. 48 'Long queues ...' Montague op. cit. (p. 90).
p. 48 'September Holocaust' *The Dogs Bulletin*, November 1939 (p. 2).
p. 48 'War came on a Sunday' Smith, *Blue Cross at War* (p. 44).
p. 48 'kitten that had been ...' Miss Audrey Neck, 10 October 1939 (M-O Diarist D5383).
p. 49 'the most dreadful ...' *Dog World*, 3 November 1939 (p. 734).

p. 49 'a queue nearly half ...' *The Animals' Defender*, November 1939 (p. 57).
p. 50 'Staff pleaded ...' Smith op. cit. (p. 44).
p. 50 'imprisoned cats ...' 'Cats Left in Empty Houses', *The Times*, 25 October 1939 (p. 2).
p. 50 'You know what they're doing' Charman, *Outbreak 1939* (p. 355).
p. 50 'Over 500 school animals ...' *The Animal World*, October 1939 (p. 1).
p. 51 'experimental animals ...' Our Dumb Friends' League Report for 1939 (p. 18).
p. 51 'The sound of gunfire ...' *The Veterinary Record*, 16 September 1939 (p. 115).
p. 51 'Animal Defence House' Lind-af-Hageby op. cit. (p. 14).
p. 51 'collect the dogs ...' *The Dog World*, 6 October 1939 (p. 640).
p. 52 'It took a great deal ...' *The Dog World*, ibid.
p. 53 'I would like you to know' Douglas, *Chronicles of Ferne* (p. 19).

Chapter 5: Keep Calm ...

p. 54 'In ever-loving memory ...' *Tail-Wagger Magazine*, October 1939 (p. 339).
p. 54 'they could not allow ...' (TNA MEPO 2/6748).
p. 54 'on the spot directions ...' ibid.
p. 55 'had in a short space ...' 19 September 1939 (TNA HO 144/21418).
p. 55 'It is something ...' TNA MEPO 2/6748
p. 55 'It would be advantageous ...' ibid.
p. 55 'we can only use our funds ...' 18 January 1940 (TNA HO 186/1418).
p. 56 'Do nothing in a panic!' *Fur and Feather*, 8 September 1939 (p. 149).
p. 56 'Every breeder in the country ...' ibid.
p. 57 'To keep rabbits ...' *The Smallholder*, 23 September 1939 (p. 1).
p. 57 'This is not just a nod to those fanciers ...' *Fur and Feather*, 8 September 1939 (p. 149).
p. 57 'We MUST strive to keep the Cat' *Fur and Feather*, 15 September 1939.
p. 57 'There must be no truce ...' *The Cat*, October 1939 (p. 1).
p. 57 'indifferent, bad and nervy ...' *The Cat*, September 1939 (p. 1).

p. 58 'If the plebeian city puss …' *The Cat*, September 1940 (p. 140).
p. 58 'Those who accompanied …' ibid. (p. 141).
p. 58 'There is a list of animal lovers …' *ARP Journal*, December 1939 (p. 15).
p. 59 'world-renowned authority …' *The Dog World*, 8 September 1939 (p. 598).
p. 59 'Dogdom and the War' *The Dog World*, 22 September 1939 (p. 616).
p. 59 'I suppose I shall carry on …' ibid.
p. 59 'England breeds …' ibid.
p. 59 'I am determined …' ibid.
p. 60 'It is early days …' ibid.
p. 60 'luckily a very nice lot …' *Our Dogs*, 29 September 1939 (p. 753).

Chapter 6: … and Carry a White Pekingese

p. 61 'A couple of hundred dogs …' Mary Golightly, *The Dog World*, 6 October 1939 (p. 640).
p. 61 'two hundred cats …' Noney Fleming, *The Dog World*, 27 December 1940 (p. 1226).
p. 61 'The secretary at Ferne …' *The Dog World*, 6 October 1939 (p. 640).
p. 62 'in batches of 10 to 20 …' Douglas, op. cit. (p. 25).
p. 63 'The evacuated cats …' ibid. (p. 62).
p. 63 'the dedication of Miss Dukie' Hageby, op. cit. (p. 29).
p. 63 '200 evacuee dogs' *Daily Mail*, 1 December 1939 (p. 7).
p. 63 'white dogs …' *The Dog World*, 29 September 1939 (p. 614).
p. 63 'Carry a white Pekingese …' Turner, *The Phoney War* (p. 70).
p. 64 'Lady Hannon …' Smith, op. cit. (p. 46).
p. 64 'a white coat for your dog …' *Daily Mail*, 29 September 1939 (p. 7).
p. 64 'Lustre Lead …' (M-OA T C 79).
p. 64 'walk in the dark …' *The Cat*, October 1939 (p. 3).
p. 64 'We should all know …' ibid.
p. 65 'Feeding Dogs in Wartime' Bob Martin Company brochure in M-OA T C 79.
p. 65 'In practice you will find ' 'Feeding Dogs', NCDL pamphlet in ibid.
p. 65 'Mother Hubbard …' 'One Woman and Her Dog', *Daily Mirror*, 15 February 1940 (p. 17).

p. 65 'a good solid pudding' 'War-Time Menus for Cats', *The Cat*, October 1939 (p. 17).
p. 65 'What Is He Going …' *Daily Mail*, 20 October 1939 (p. 7).
p. 66 'animal wardens on every street …' (TNA HO 45/18150).
p. 67 'Guard No. 1', *Manchester Guardian*, 13 January 1941 (p. 3).
p. 67 'Animal Guards are Needed Now!' *Animal and Zoo News*, Dec 1939 (p. 3).
p. 67 'Do not have your pet destroyed …' (TNA HO 186/1417).
p. 67 'A great army of National …' NARPAC Bulletin No. 2, December 1939 M-OA T C 79.
p. 68 'Offers of help poured in …' Stone, *Salute to the National Animal Guard*, 1945 (p. 14).
p. 68 'For the duration of the war …' NARPAC Bulletin No. 2.
p. 69 'The PDSA has placed …' ibid.
p. 69 'Who will help …' (TNA HO 186/1418).
p. 69 'This country has gone to war …' ibid.
p. 69 'I assure you …' ibid.
p. 69 'The officers and men at a wireless …' *The Dog World*, 6 October 1939 (p. 640).
p. 70 'Detached posts such as RDF …' (TNA WO 199/416).
p. 70 'I could get a lot of material …' ibid.
p. 70 'so fierce it was a danger …' ibid.
p. 70 'it could not be relied upon …' ibid.

Chapter 7: Hunting Must Continue!

p. 71 'Governesses' carts …' Graves, *Great Days*, Hutchinson, London, 1944 (p. 52).
p. 72 'The King has permitted …' (TNA HO 186/1418).
p. 72 'cubbing should take place …' (TNA MAF 79/8).
p. 72 'Hunting must continue!' *Horse & Hound*, 22 September 1939 (p. 1339).
p. 73 'War or no war …' Turner, op. cit. (p. 162).
p. 73 'war is upon us …' *The Field*, 23 September 1939 (p. 682).
p. 73 'No brave scarlet …' *The Field*, 14 December 1939 (p. 974).
p. 73 'colonial or dominion troops …' (TNA MAF 35/494).
p. 74 'Soldiers and airmen …' ibid.
p. 74 'Major Montacute …' Holloway, Hounds, Hares and Foxes of Larkhill, Larkhill, Royal Artillery Hunt, 1992.

p. 74 'managed to smuggle ...' Obituary of Lt-Col Frederick Edmeades, *Daily Telegraph*, 1 May 2001.
p. 74 'should take advantage ...' Hansard, 2 October 1939.
p. 74 'Where would the eighteen ...' (TNA MAF 79/8).
p. 75 'What I hear from ...' (TNA MAF 79/8).
p. 75 'If he had his way ...' Lord Claud Hamilton, *The Times*, 13 March 1918 (p. 3).
p. 75 'It is too early yet ...' Hansard, 19 October 1939.
p. 76 'Those clamouring for ...' Lind-af-Hageby op. cit. (p. 115).
p. 77 'Shooting men can get ...' *The Field*, 10 February 1940 (p. 204).
p. 77 'We appeal to every ...' *Eggs*, 6 September 1939 (p. 281).
p. 77 'Rabbits are not ...' Orwell diaries, 28 September 1939 (http://orwelldiaries.wordpress.com/28-9-39/).
p. 78 'Each and every goat ...' *The Goat*, No. 1, 1 November 1939.
p. 78 'Bees as messengers ...' *PDSA News*, June 1943 (p. 10).

Chapter 8: Wolves Not Welcome

p. 80 'Submerged in their wallows ...' Ackerman, *The Zookeeper's Wife* (p. 62).
p. 81 'frightened elephant ...' Heck, *Animals – My Adventure* (p. 91).
p. 83 'When ordinary Germans ...' *Animal and Zoo Magazine*, November 1939 (p. 6).
p. 83 'Orders were given for all beasts ...' Heck, op. cit. (p. 91).
p. 84 'a few of the bigger chimpanzees ...' *Animal and Zoo Magazine*, November 1939 (p. 7).
p. 84 'the animals will enjoy ...' *Animal and Zoo Magazine*, November 1939 (p. 2).
p. 84 'only a few redundant ...' *The Times*, 14 October 1939 (p. 4).

PART TWO: THE MINISTRY OF PETS

Chapter 9: Pets Get the Blame

p. 91 'grave concern ...' *The Times*, 1 January 1940 (p. 14).
p. 92 'Even dogs ...' *The Times*, 8 January 1940 (p. 12).
p. 92 'It was always after ...' *Western Morning News*, 23 February 1940 (p. 3).

p. 92 'The Market was also …' ibid.
p. 92 'Although not affected …' ibid.
p. 92 'The black-out had …' *Western Morning News*, 5 January 1940 (p. 3).
p. 93 'This curious war …' *The Animals' Defender*, December 1939 t/s in M-OA T C 79.
p. 93 'The aim of …' *The Times*, 6 January 1940 (p. 8).
p. 93 'NARPAC Caught …' *The Animals' Defender*, March 1940 (p. 86).
p. 94 'highly inflammable …' 4 March 1940 (TNA HO 186/1418).
p. 94 'looking for its master …' *Daily Mail*, 14 February 1940 (p. 7).
p. 94 'five white ferrets …' Our Dumb Friends' League Report for 1940 (p. 58).
p. 94 'His cat, Whiskers …' *PDSA News*, March 1940 (p. 1).
p. 95 'but he arrived safely …' *Daily Mail*, 28 March 1940 (p. 7).
p. 95 'a frightened tabby …' *Our Dogs*, June 1942 (p. 752).
p. 96 'tragic and unspeakably heroic …' *The Animals' Defender*, March 1940 (p. 86).
p. 96 'the faithful little …' ibid.
p. 98 'Wartime Aids …' (pp. 13–18) 'Wartime Food for Dogs and Cats' (TNA HO 144/21418).
p. 98 'This country has not reached …' *The Animal World*, March 1941 (p. 21).
p. 99 'the majestic rise of the vegetarian …' Hageby op. cit. (p. 38).
p. 100 'Adolf' the Aardvark *Evening Standard*, 23 September 1942 (ZSL Newscuttings vol. 37).
p. 100 'dreadful people …' *Tail-Wagger Magazine*, December 1939 (p. 372).
p. 100 'would fall on the lower classes …' 'The Feeding of Dogs' (para. 7), 22 July 1941 (MAF 84/61).
p. 101 'Some will take pieces …' 'Tortoises in Wartime', *The Animal World*, August 1940 (p. 68).
p. 101 'Away goes our last source of imported …' Turner op. cit. (p. 123).
p. 101 'Our hobby …' ibid.
p. 102 'He was treated with …' Fred Gluckstein (www.winstonchurchill.org-churchills-feline-menagerie).

p. 102 'Nelson will follow ...' *Washington Post*, 7 June 1940 (p. 8).
p. 102 'became the pet of ...' 4 March 1965 (TNA BA 20/39).
p. 103 'Mr J. Umbo ...' (TNA CAB 150/7).
p. 103 fn 'allergic to cats' Nerin E. Gun, *Eva Braun: Hitler's Mistress* L. Frewin, London, 1969 (p. 212).

Chapter 10: The Dunkirk Dogs

p. 104 'We passed through village ...' 2nd Lt E. J. Haywood writing in 1946. (www.worcestershireregiment.com).
p. 105 'Shooting the dogs ...' Edward Oates in 2010 (ww2history.com/testimony/Western/British_soldier_Dunkirk).
p. 105 'pitiful procession of refugees ...' *The Animal World*, November 1940 (p. 77).
p. 105 'large numbers of dogs ...' Moss op. cit. (p. 113).
p. 106 'knowing with ...' ibid.
p. 106 'Boxer ...' Bourne, *They Also Serve* (p. 82).
p. 106 'There was a little dog ...' Ordinary Seaman Stanley Allen, IWM Interview 6825 (http://www.iwm.org.uk/collections/item/object/80006641).
p. 106 'evacuated a very special person ...' Stockport Libraries, (http://www.bbc.co.uk/history/ww2peopleswar/stories/16/a2314216.shtml).
p. 106 'One of the little ...' Our Dumb Friends' League Report for 1940.
p. 106 'Men of the French ...' ibid.
p. 107 'a large, fierce-looking ...' *Tail-Wagger Magazine*, July 1940.
p. 107 'returned ...' *The Cat*, June 1940 (p. 1).
p. 107 'All along the Thames ...' RSPCA Report for 1940 (p. 10).
p. 108 'after Dunkirk was ...' IWM Allen, op. cit.
p. 108 'Where the owners ...' RSPCA Report for 1940 (p. 10).
p. 108 'there were many ...' *The Animal World*, July 1940 (p. 59).
p. 108 'been segregated ...' *The Animal World*, July 1940 (p. 60).
p. 109 'Many dogs attached ...' Our Dumb Friends' League Report for 1940 (p. 30).
p. 109 'there was not just ...' ibid.
p. 109 'not a little difficulty ...' RSPCA Report for 1940 (p. 10).
p. 110 'Fortnum ...' *PDSA News*, June 1942 (p. 5).
p. 110 'the last dog from Dunkirk ...' RSPCA Report for 1940 (p. 10).

p. 110 'Nora …' ibid.
p. 111 'The Dogs from Dunkirk …' *The Dogs Bulletin*, No. 110, summer 1940.

Chapter 11: Three Million Dogs to Die

p. 112 'felt they had nothing …' 10 June 1940 (TNA HO 186/2075).
p. 112 'to protect the public …' ibid.
p. 113 'The following towns …' (TNA HO 186/350).
p. 113 'had arranged with …' Kean, *Animal Rights* (p. 192).
p. 113 'beloved ginger cat …' Diana Mosley, *Loved Ones*, 1985 (p. 192).
p. 114 'The man who …' Colin Perry, *Boy in the Blitz*, 1980 (p. 13).
p. 114 'animals' true friend …' *Reichtierschutzblatt*, May 1940 (http://elib.tiho-hannover.de/dissertations/bergb_ws08.pdf).
p. 114 'a French battlefield stray …' ibid. February 1941.
p. 114 'fundamental rottenness …' Anger Campaign (p. 27, TNA FO 898/5).
p. 115 'Three Million Dogs', *The Times*, 13 June 1940 (p. 6).
p. 115 'was more upset …' Martina Corfe (Mass-Observation diarist D5285).
p. 115 'a newsboy …' Helena Datz (Mass-Observation diarist D5294).
p. 116 'Can it be true?' Thelma Gray, *The Dog World*, 21 June 1940 (p. 577).
p. 116 'These two Nazi officials …' ibid.
p. 116 'It must have been with …' *The Dog World*, 21 June 1940 (p. 577).
p. 116 'They have shared …' ibid.
p. 116 'glycerine and the fertilisers …' 24 June 1940 (TNA HO 186/1418).
p. 117 'training experiments …' Hansard, 21 May 1940.
p. 117 'British Government …' *The Dog World*, 3 May 1940 (p. 414).
p. 118 'A dog trained to do …' Richardson, *British War Dogs* (p. 287).
p. 118 'When war broke out …' 11 June 1940 (TNA WO 199/416).
p. 119 'whether he was aware …' Hansard, 21 May 1940.
p. 119 'experimental work …' *The Dog World*, 31 May 1940 (p. 509).
p. 119 'One could surely coach …' *The Dog World*, 5 July 1940 (p. 625).

p. 119 'In my opinion ...' *Our Dogs*, 5 July 1940 (p. 810).
p. 119 'last remaining remount ...' *The Times*, 10 August 1940 (p. 2).
p. 120 'The green plover ...' Turner, op. cit. (p. 249).
p. 120 'Dogs and the Invader ...' NCDL pamphlet No. 469 Ballard, 'A Dog is For Life' (p. 68).
p. 120 'Poultry and rabbits ...' 8 August 1940 (TNA HO 186/1224).
p. 120 'for animal control ...' War Cabinet Home Policy Committee, 29 June 1940 (TNA CAB 75/8 6398).
p. 121 'A room for cats ...' 16 July 1940 (TNA AN 2/29).
p. 121 'personally enlisting ...' *The Veterinary Record*, 13 July 1940 (p. 519).
p. 122 'The worst time ...' *PDSA News*, June 1942 (p. 3).
p. 122 'veterinary officers ...' 13 July 1940 (TNA HO 186/1419).
p. 123 'As there was no room ...' *The War Illustrated*, 19 July 1940 (p. 40).
p. 123 'All our animals' Evacuation from Guernsey' by Betty Hervey (http://www.bbc.co.uk/history/ww2peopleswar/stories/39/a2871939.shtml).
p. 123 'the squealings of ...' *Our Dogs*, 5 July 1940 (p. 810).
p. 124 'We must bestir ourselves ...' ibid.
p. 124 'Fears are expressed ...' *The Dogs Bulletin*, No. 116, summer 1940 (p. 5).
p. 124 'We rejoice that American ...' *Popular Dogs*, July 1940 (www.adpef.org/breed_development/first.../popular_dogs_July-1940).

Chapter 12: No Cat Owner Need Worry

p. 125 'Penal Offence' *Gloucestershire Echo*, 6 August 1940 (p. 6).
p. 126 'nature of a warning ...' ibid.
p. 126 'sheltered sunny positions ...' (TNA MAF 84/57 7041).
p. 126 'Mickey' Longmate, op. cit. (p. 222).
p. 127 'wasting human food ...' (TNA MAF 84/57 6945 5982).
p. 127 'It does not seem possible ...' *The Animal World*, September 1940 (p. 79).
p. 127 'is it antipatriotic ...?' *The Cat*, September 1940 (p. 111).
p. 128 'people have found ...' *Daily Mirror*, 8 January 1940 (p. 7).

p. 128 'Feeding Dogs and Cats', The Duchess of Hamilton and Louise Lind-af-Hageby, *The Times*, 12 August 1940 (p. 6).
p. 129 'many dogs are voluntary vegetarians …' ibid.
p. 129 'the people's food' *Farmers Weekly*, 9 August 1940 (p. 13).
p. 129 'Stock casualties in Nazi raids' *Farmers Weekly*, 23 August 1940 (p. 17).
p. 130 'There has been no …' ibid.
p. 130 'Nazi bombs caused low buttermilk' *Farmers Weekly*, 30 August 1940 (p. 17).
p. 131 'Dogs of Britain Fighter Fund' *Our Dogs*, 23 August 1940 (p. 983).
p. 131 'Lady Johnson-Ferguson' (TNA MAF 84/57).
p. 131 'Two glorious days …' 16 August 1940 (p. 273 http://orwelldiaries.wordpress.com/2010/08/16/16-8-40).

Chapter 13: Blitz Pets

p. 133 'Item 1 – The Restriction' 4 September 1940 (TNA 128/578).
p. 133 'dog racing and the maintenance of zoos …' ibid.
p. 133 'Any drastic steps …' ibid.
p. 133 'There should be no action …' ibid.
p. 134 'Germany is presently …' (TNA WO 186/350).
p. 134 'You must assume invasion …' (TNA WO 186/350).
p. 135 'Tonight I have been registering …' 29 July 1940 (M-O D5414).
p. 135 'Dogs left behind …' 18 September 1940 TNA MEPO 2/6478.
p. 136 'again we had to open …' 'Our War Work', *PDSA News*, June 1943 (p. 4).
p. 137 'His home was a mass …' Bourne, op. cit. (p. 56).
p. 137 'Cats have some sixth …' Our Dumb Friends' League Report for 1940 (p. 27).
p. 137 'but six cats …' ibid. (p. 76).
p. 138 'An old man stopped me …' Lind-af-Hageby, op. cit. (p. 25).
p. 138 'Dog belonging to Mrs H …' ibid. (p. 26).
p. 139 'Trixie …' ibid.
p. 141 '19,864 Cats', Our Dumb Friends' League Report for 1940 (p. 60).
p. 141 'after an absence of three …' ibid.
p. 141 'deserted animals …' ibid.

p. 141 'neither of them is afraid ...' *The Dogs Bulletin*, summer 1941 (p. 7).
p. 141 'Animal pets are frequently ...' M-O Summary of Situation in Shoreditch, 27 September 1940 (M-OA T C 79).
p. 142 'the shock of emerging ...' Longmate, op. cit. (p. 224).
p. 142 'Many piteous tales of cats ...' Our Dumb Friends' League Report for 1940 (p. 54).
p. 142 'to find new homes ...' Our Dumb Friends' League Report for 1940 (p. 57).
p. 143 'A monkey ...' ibid.
p. 143 'becoming abusive ...' 'Dogs in Air-Raid Shelters', *Manchester Guardian*, 4 October 1940 (p. 10).
p. 143 'because of the crass ...' *The Dogs Bulletin*, No. 117, Christmas 1940 (p. 5).
p. 144 'This air-raid dog shelter ...' ibid.
p. 144 'a conspicuous and attractive ...' Lind-af-Hageby, op. cit. (p. 24).
p. 144 'Everywhere the poor folk ...' ibid.
p. 145 'He is all I have now ...' *PDSA News*, January 1941 (p. 2).
p. 145 'Such is the cleanliness ...' Our Dumb Friends' League Report for 1940 (p. 55).
p. 145 'She had also ...' ibid.
p. 146 'sentimental aspects ...' (TNA MAF 126/25 7072).
p. 146 'wholly won over ...' Longmate, op. cit. (p. 231).
p. 147 'At the end of their ...' Stafford Library, Kent (http://www.bbc.co.uk/history/ww2peopleswar/stories/02/a8435702.shtml).
p. 147 'Chickens were killed ...' Mr Peter Johnson (http://www.1900s.org.uk/1940s50s-livestock.htm.
p. 148 'Then my mother got these rabbits ...' (www.yorkarchaeology.co.uk/ww2/rations1.htm).
p. 148 'One day one was killed ...' Longmate, op. cit. (p. 231).
p. 148 'never skinned a rabbit ...' ibid.
p. 148 'a large, fat, white rabbit ...' ibid.
p. 148 'My mum ...' (http://www.sarfend.co.uk/cgi-bin/forum).

Chapter 14: The Comfort of Pets

p. 149 'Readers may conclude ...' Anon t/s in RSPCA archive.
p. 149 'The writer has an old ginger cat ...' ibid.

p. 150 'I am suffering …' *The Dogs Bulletin*, No. 117, Christmas 1940 (p. 7).
p. 150 'which will be spun …' *The Dog World*, 11 October 1940 (p. 919).
p. 151 'kitten seemed terrified …' 13 November 1940 (M-O Diarist D5428).
p. 151 'went to a house soon …' Moss, op. cit. (p. 42).
p. 151 'a very frightened and bedraggled tabby …' ibid (p. 41).
p. 152 'carried her kittens one by one …' ibid.
p. 152 'Underneath her …' *Daily Mirror*, 8 October 1940 (p. 7).
p. 152 'Beauty', *PDSA News*, April 1941 (p. 2).
p. 152 'Her sharp senses will …' *Dog World Annual*, 1941.
p. 153 'No one knew …' Bourne, op. cit. (pp. 158–9).
p. 153 'modest little dog …' Our Dumb Friends' League Report for 1940 (p. 28).
p. 153 'Peggy … terrier of sorts' Moss, op. cit. (p. 42).

Chapter 15: How is Your Pet Reacting?

p. 154 'How is your dog reacting?' *The Dogs Bulletin*, No. 117, Christmas 1940 (p. 6).
p. 154 fn 'the demoralising wail …' Churchill Archives (CHAR 20/10/13-14).
p. 154 'Cats are comparatively …' *PDSA News*, October 1940 (p. 8).
p. 155 'My bees are frightened …' *Bee Craft*, October 1940 (p. 185).
p. 155 'a bomb made …' *Bee Craft*, November 1940 (p. 199).
p. 155 'crept under a sofa …' *ARP News*, October 1941 (p. 25).
p. 155 'A pack of startled …' ibid.
p. 155 'abandoned it …' ibid.
p. 155 'All the women …' George Orwell Diary, 17 September 1940 (p. 292) (http://orwelldiaries.wordpress.com/17-9-40).
p. 155 'also like this during raids …' ibid.
p. 156 'Some people have …' *Tamworth Herald*, 20 July 1940 (p. 2).
p. 156 'a really well trained …' *ARP News*, May 1941 (p. 6).
p. 156 'A timid and excited …' The Animal World, November 1940 (p. 92).
p. 156 'Animals are very sensible …' Our Dumb Friends' League Report for 1940 (p. 63).
p. 156 'the engine became …' *The Dogs Bulletin*, No. 117, Christmas Number (p. 6).

p. 157 'My own Collie ...' *ARP News*, May 1941 (p. 6).
p. 157 'Badly trained dogs ...' ibid.
p. 157 'At the alert ...' *Animals and Air Raids: A Study in Reactions Times*, 14 October 1940 (p. 7).
p. 157 'miserably frightened ...' ibid.
p. 157 'Some parrots ...' ibid.
p. 157 'scattering downwards ...' ibid.
p. 158 '38 incendiary bombs' ZSL Occurrences Book, 27 September 1940.
p. 158 'Twenty-four rhesus monkeys ...' ibid.
p. 159 'By a miracle ...' 'Zoo Animals' Calmness in Air Raids', *The Times*, 2 December 1940 (p. 2).
p. 159 'A Zebra was liberated ...' ibid.
p. 159 'unease in the country ...' 16 October 1940 (TNA HO 144/21418).

PART THREE: PETS SEE IT THROUGH

Chapter 16: The Perils of NARPAC

p. 165 'sorry it had come to this ...' (TNA HO 186/2075).
p. 166 'suspicious of everyone ...' ibid.
p. 166 'wallowing in publicity ...' ibid.
p. 166 'Their leaders are without conscience ...' ibid.
p. 166 'extremists ...' ibid.
p. 166 'normal flow of legacies ...' ibid.
p. 167 'NARPAC has no legal existence ...' ibid.
p. 167 'not interested in any of it ...' ibid.
p. 168 'had taken over the staff at Harrison, Barber & Co. ...' ibid.
p. 169 'Harrison, Barber & Co.', *The Times*, 20 March 1941 (p. 8).
p. 169 'could only dispose of' (TNA MEPO 2/6589).
p. 169 'climbed out of ...' ibid.
p. 169 'finding homes' ibid.
p. 169 'prevent an undertaker ...' 8 November 1940 (TNA HO 186/2075).
p. 170 'Instead of going mad ...' 'Dogs and Cats Take Raids Calmly', *The New York Times*, 26 November 1940 (p. 3).
p. 171 'in which certain ...' (TNA HO 186/2075).
p. 172 'That plaintive cry ...' *PDSA News*, December 1940 (p. 8).

p. 172 'I have rescued hundreds …' *Daily Mirror*, 3 December 1940 (p. 11).
p. 172 'At midnight, a goat …' *Manchester Guardian*, 13 January 1941 (p. 3).
p. 173 'wild, starving cats …' *Manchester Guardian*, 11 January 1941 (p. 10).
p. 173 'these cats are wild and elusive …' ibid.
p. 173 'the kindly owner of a local …' *The Dog World*, 27 December 1940.
p. 173 'Puss placed her four …' *The Cat*, June 1941 (p. 87).
p. 174 'a huge black Tom …' ibid.

Chapter 17: Non-essential Animals

p. 176 'Throughout the country …' *Manchester Guardian*, 7 January 1941 (p. 3).
p. 176 'Dusty, a cat from …' ibid.
p. 176 'un-tethered the horses …' LAPAVS News-Sheet, August 1941 (p. 1).
p. 177 'where farmers have found …' *Manchester Guardian*, 7 January 1941 (p. 3).
p. 177 'killed two good plough horses …' ibid.
p. 178 'Owners of urban …' *The Times*, 10 January 1941 (p. 2).
p. 178 'I fancy these leaflets …' *Manchester Guardian*, 7 January 1941 (p. 3).
p. 179 'Five couple of hounds …' Turner, op. cit. (p. 164).
p. 179 'made into glue …' ibid.
p. 179 'Courtenay Tracey …' 9 January 1941 (TNA MAF 79/8).
p. 179 '7,800 hounds have been …' (TNA MAF 79/8).
p. 180 'We suggest that hunting …' (TNA MAF 128/578).
p. 180 'I emphasise the political …' ibid.
p. 181 'when all imports …' (TNA MAF 84/57).
p. 181 'the sole source …' Hansard, 2 April 1941.
p. 181 'Why is the hunt allowed …' Mrs Elsie Barnes, 25 January 1941 (TNA MAF 79/8).
p. 181 'Dogs and cats must subsist …' (TNA MAF 128/578).
p. 181 'I think I am right …' (TNA MAF 85/87).
p. 182 'seagulls, city pigeons etc.' 26 March 1941 (TNA MAF 84/57).
p. 182 'An increasing number …' (M-O SxMOA1/2/79/1/A/1).
p. 182 '17,347 dogs destroyed' (TNA MEPO 2/6957).

p. 182 'living in the paraffin shed' Jenkins, *A Home of Their Own* (p. 175).
p. 183 'expert electrocutionist …' 28 May 1941 Mrs M. Witherow to C. Pulling (TNA MEPO 2/6589).
p. 183 'There appears to be …' (TNA MAF 84/57).
p. 183 'a cat equilibrium …' (TNA MAF 84/57 5857).
p. 183 'It seems possible that an attempt …' *The Cat*, February 1941 (p. 1).
p. 184 '10–12 circuses …' (TNA MAF 84/57).
p. 184 'Watson's Fox Terriers' *Tail-Wagger Magazine*, March 1943 (p. 45).
p. 184 'As nations allow animals …' Performing Animals (TNA HO 45/20552).
p. 185 'clearly indicated fear …' Performing Animals (TNA HO 45/20552).
p. 185 'animals, dingoes …' Our Dumb Friends' League Report for 1940 (p. 37).
p. 185 'unless this war …' *The Times*, 11 March 1941 (p. 5).
p. 186 'Millions of people …' (TNA MAF 79/8).
p. 186 'my section of …' George Orwell diaries, 25 April 1941 (pp. 306–7, http://orwelldiaries.wordpress.com).
p. 186 'Lord Woolton's promise …' *The Evening News*, 25 March 1941 (p. 1)
p. 186 'feeding of pigeons …' ibid.
p. 187 'There should no …' 'Animal Feeding Stuffs. Dogs and Cats', 1 April 1941 (TNA MAF 128/578 5684).
p. 187 'Interfering with …' 22 July 1941 (TNA MAF 84/61).
p. 187 'doing away with all …' (TNA MAF 84/57 29 April 1941).
p. 187 'this applies to …' Hansard, 2 April 1941.
p. 188 'Poultry-keepers …' *Glasgow Herald*, 1 March 1941 (p. 1).
p. 188 'complete denial …' Hansard, 30 April 1941.
p. 188 'far-reaching motive …' *The Cat*, February 1941 (p. 66).
p. 188 'It is difficult to tell …' ibid.
p. 189 'prepared to go to …' ibid. (p. 67).

Chapter 18: Pets Under Fire

p. 190 'There was this scheme …' NCDL, 17 March 1941 (M-OA T C 79).
p. 191 'In a case heard …' *The Cat*, August 1941 (p. 122).

p. 191 'The rescue squad …' *Western Morning News*, 24 June 1941 in *LAPAVS News-Sheet*, August 1941.
p. 192 'many cats destroyed …' ibid.
p. 192 'Housewives take him …' LAPAVS News-Sheet, May 1941 (p. 1).
p. 192 'Blackie' *Daily Mirror*, 6 November 1941 (p. 4).
p. 192 'A cat buried for …' *PDSA News*, June 1941 (p. 2).
p. 193 'Four days later …' ibid.
p. 193 'cats, dogs, chickens …' ibid.
p. 193 'It is worth mentioning that …' *Our Dogs*, 14 June 1941 (p. 752).
p. 194 'I noticed a large …' Smith, op. cit. (p. 49).
p. 194 'where she worked …' ibid. (p. 63).
p. 195 'Florence Nightingale' *Daily Mirror*, 20 September 1943 (p. 4).
p. 195 'a friend in London …' *The Cat*, February 1941 (p. 62).
p. 195 'animal lover' Smith op. cit. (p. 63).
p. 196 'Someone asked me …' Lind-af-Hageby op. cit. (p. 24).
p. 196 '1,400 pets, mostly cats …' RSPCA Report for 1941 (p. 16).
p. 196 'In one East End …' Our Dumb Friends' League Report for 1940 (p. 57).
p. 196 'Cats are the most …' *The Veterinary Record*, 5 July 1941 (p. 390).
p. 197 'has fed hundreds …' *Daily Mirror*, 2 November 1940 (p. 3).
p. 197 'rather than see the pack cease …' *Daily Mail*, 25 October 1941.
p. 198 'I remember Clydesdale …' John Maude, 11 January 1941 (TNA KV 2/1684).
p. 199 'The Duke's mother …' ibid.
p. 199 'We had lots of voluntary …' Smith, op. cit. (p. 45).

Chapter 19: Wanted – Dog Heroes

p. 201 'followed by another cat …' *Birmingham Mail*, 19 April 1941 quoted in *The Cat*, July 1941 (p. 110).
p. 201 'a small cafe, with a nice cat …' *The Cat*, September 1941 (p. 4).
p. 201 'My own two cats …' ibid.
p. 202 'sprang up, ran full tilt …' *The Star*, 28 March 1941, quoted in *LAPAVS News-Sheet*, May 1941 (p. 1).

p. 202 'The dog is a friend ...' *LAPAVS News-Sheet*, May 1941 (p. 1).
p. 202 'Once there they ...' The Siamese Cat Club News-Sheet, quoted in *The Cat*, October 1940 (p. 14).
p. 202 'that knew the difference ...' *The Cat*, June 1941 (p. 87).
p. 203 'The large majority of beasts ...' 'War Time Reactions of Cats', *The Cat*, February 1941 (p. 88).
p. 203 'It is, of course, difficult ...' ibid. (p. 89).
p. 203 'it seemed incredible ...' Lutz, op. cit. (p. 93).
p. 203 'soon realising ...' ibid.
p. 204 'When Nazi bombers ...' *The Dogs Bulletin*, No. 117, Christmas 1940 (p. 7).
p. 204 fn 'Ome ...' *The Dog World*, 27 October 1939 (p. 709).
p. 205 'Throughout the country ...' *The Veterinary Record*, 5 July 1941 (p. 389).
p. 205 'Cats bombed out ...' ibid.
p. 205 'Cattle appear to take ...' ibid.
p. 205 'slept sonorously ...' ibid.
p. 206 'hither and thither ...' ibid.
p. 206 'The herding instincts ...' Clifford W. Greatorex, *ARP News*, October 1941 (p. 25).
p. 206 'where he stayed ...' ibid.
p. 206 'ordinary men and women ...' letter from 'Not an Owner of Walthamstow' in *News Chronicle*, 20 June 1941 (p. 2).
p. 206 'increased pro-doggism ...' Mass-Observation File Report 804, 'Dogs in London'.
p. 206 'mongrels took it ...' ibid.
p. 207 'When I leave him ...' ibid.
p. 207 'I have been alone ...' ibid.
p. 207 'Nip', 'Bill', 'Bonzo', ibid.
p. 207 'there should be more done ...' ibid.

Chapter 20: Pets on the Offensive

p. 208 'It would be impossible ...' (TNA WO 199/416).
p. 208 'The War Office invites ...' *The Times*, 6 May 1941 (p. 2).
p. 209 'Well, pals, here's how ...' 'The War and Us Dogs by Rex', *Tail-Wagger Magazine*, September 1944 (p. 175).
p. 209 'will answer to Oi!' *PDSA News*, August 1941 (p. 5).

p. 210 'full of dogs …' Montague, op. cit. (p. 78).
p. 211 'smuggled out in boxes …' ibid.
p. 211 'one little dog …' ibid.
p. 211 'The platoon did not return …' ibid. (p. 79).
p. 211 'babies were painlessly …' Bourne, op. cit. (p. 75).
p. 211 'The grief-stricken owner …' Montague, op. cit. (p. 81).
p. 212 'dogs circling …' *Horse & Hound*, 1 January 1943 (p. 15).
p. 213 'I had lunch with …' Colville, *The Fringes of Power*, vol. 1 (p. 468).
p. 213 'with no statutory funding …' (TNA HO 186/2075).
p. 214 'Although we could still …' ibid.
p. 214 'I think it's made him …' (SxMOA1/1/6/7/35).
p. 214 'He kicks up an awful row …' ibid.
p. 214 'so well mannered …' ibid.
p. 214 'I used to have a dog …' ibid.
p. 215 'I adored him …' ibid.
p. 215 'When a country is …' (M-OA TC79).
p. 215 'Rush to shop to get fish …' 3 June 1944, Anon M-O diarist D5444.
p. 216 'There's that notice up saying …' (M-OA TC79).
p. 216 'in reply to inquirers …' *The Veterinary Record*, 26 July 1941 (p. 441).
p. 216 'any interference with dogs and pets …' (TNA MAF 84/61).
p. 217 'the use of propaganda in the case of cats …' 29 July 1941 (TNA MAF 84/61).
p. 218 'the biggest cat …' *Manchester Evening Chronicle*, 18 July 1941 (reported in *The Cat*, August 1941 (p. 122)).
p. 218 'you seem to think …' 7 August 1941 (TNA MAF 48/757).
p. 219 'Wayward cats …' *Dig for Victory News*, No. 68, 13 June 1941 in TNA MAF 48/757.
p. 219 'My Lord – I have a large bed of onions …' (TNA MAF 48/757 6255).
p. 219 'peculiar black markings' *PDSA News*, October 1941 (p. 10).
p. 220 'domestic cats live …' Schwangart, *Vom Recht der Katze* (p. 15).
p. 220 'Nelson is the bravest …' Reynolds, *All About Winston Churchill* (p. 148).
p. 221 'faux pas …' *The Cat*, December 1941 (p. 33).
p. 222 'There seems to be little …' (MAF 48/757 6468).
p. 222 'being somewhat …' ibid.

p. 223 'All exhibitors and visitors ...' Worcester Dog Show (M-OA TC79/1/F).
p. 223 'Many wore riding ...' ibid.
p. 223 'even with a war on ...' *PDSA News*, October 1941 (p. 3).
p. 224 'As the war goes on ...' *The Times*, 16 December 1941 (p. 2).
p. 224 'See how elastic ...' The Official Home Office Cat, 22 January 1941 (TNA HO 223/43 6413).
p. 225 'whether a small daily ration of milk ...' Hansard, 26 November 1941.
p. 225 'Limited quantities ...' ibid.
p. 225 'work of national importance ...' *The Times*, 31 December 1941 (p. 2).

Chapter 21: Nationally Important Cats

p. 226 'Lord Woolton's face can ...' *Manchester Guardian*, 1 January 1942 (p. 6).
p. 226 'extraordinarily fussy cat ...' ibid.
p. 226 'We must not be jealous ...' *The Times*, 8 January 1942 (p. 5).
p. 227 'a lenient view ...' 'Feeding Our Pets in War Time', *The Animal World*, December 1941 (p. 1).
p. 227 'Combings from almost every kind ...' *The Dogs Bulletin*, spring 1941 (p. 11).
p. 227 'rat watchers ...' (see TNA HO 196/2486).
p. 228 'Mopsy' *The Dogs Bulletin*, spring 1942 (p. 5).
p. 229 'Welsh moss ...' *The Field*, 25 January 1947 in ZSL newscuttings.
p. 229 'even stepping into the dry ...' 'Starving the Elephants: The Slaughter of Animals in Wartime Tokyo's Ueno Zoo', Frederick S. Litten (http://www.japanfocus.org/-frederick_s_-litten/3225).
p. 229 'Airedales and Alsatians ...' *The Animal World*, April 1942 (p. 1).
p. 230 'We can only use ...' *Tail-Wagger Magazine*, August 1945 (p. 145).
p. 230 'violent opposition ...' (TNA WO 199/416 7263).
p. 231 'never saw an English patrol dog ...' *The Dog World*, 3 May 1940 (p. 414).
p. 231 'useless ...' (TNA AVIA 9/15 7377).
p. 232 'He can get the necessary ...' ibid.
p. 232 'the thing we want the dog ...' ibid.

p. 232 'Plough Inn …' 'Working Dogs for the Protection of Aircraft', 8 September 1941 (TNA WO 199/2061).
p. 233 'Any dog debarred …' 9 July 1942 (TNA WO 199/416).
p. 233 'tattooing the ear flap …' 5 March 1942 (TNA WO 199/416).
p. 233 'The public will be asked to lend them …' 18 March 1942 (TNA WO32/10800).
p. 233 'the wastage will be …' ibid.
p. 234 'to make a valuable …' Draft Press Appeal (TNA WO32/10800).
p. 234 'over 6,000 owners …' *The Times*, 14 January 1944 (p. 2).
p. 234 'I would be much obliged …' (TNA WO32/10800).
p. 234 'The handler continually …' 'The Tactical Employment of War Dogs', August 1942 (WO 199/416).
p. 235 'The team of eight dogs …' 16 April 1942 (TNA AVIA 9/15).
p. 235 'that women who …' Hansard, 4 March 1942.
p. 236 'a census of cats …' ibid.
p. 236 'Are any steps …' ibid.

Chapter 22: Too Many Poodles

p. 237 'The keepers of the forest …' (TNA MAF 79/9).
p. 237 'Sanderstead and Coulsdon' 22 May 1942 (TNA MAF 79/9).
p. 238 'Jowitt, your minister …' (TNA PREM 4/2/11).
p. 238 'never seen a cattle …' (TNA MAF 100/19).
p. 238 'unfair that every officer …' (TNA MAF 100/19).
p. 239 'not want a long delay …' 27 March 1942 (TNA MAF 100/21).
p. 239 'I understand that …' ibid.
p. 239 'If there was no prosecution …' 18 July 1942 (PREM 4/2/11).
p. 239 'The customers of Albion …' *The Times*, 29 August 1942 (p. 2).
p. 240 'Goods have to be …' April 1942 (TNA MAF 84/57).
p. 240 'they are no longer …' ibid.
p. 240 'It is not the intention …' 'The Feeding of Cats and Dogs in Wartime' (TNA MAF 84/57).
p. 240 'we might cause it to be known …' 7 March 1942 (TNA MAF 84/61).
p. 241 'other than strays …' (TNA MAF 84/61).
p. 241 'of the political implications …' ibid.
p. 241 'not desirable that NARPAC …' ibid.

p. 241 'We hear of a trainload ...' Lambeth Palace Library Archbishop Temple, 18 September 1942 (vol. 1 f329-332).
p. 242 'millions of useless cats ...' cutting in TNA MAF 84/61.
p. 242 'Humanely exterminate ...' ibid.
p. 242 'One Dog per family!' ibid.
p. 242 'Too many cats ...' ibid.
p. 243 'dogs getting biscuits ...' ibid.
p. 243 'Nero, a dog the size of a horse ...' ibid.
p. 243 'while sporting dogs ...' *Daily Mirror*, 29 June 1942.
p. 243 'silly attacks ...' Lind-af-Hageby, op. cit. (p. 48).
p. 244 'They are friends ...' ibid.
p. 244 'Night Worker ...' *Gloucestershire Echo*, 10 September 1942.
p. 244 'the collecting dog ...' ibid.
p. 244 'too many poodles ...' *Farmers Weekly*, 23 October 1942.
p. 244 'we are a nation ...' ibid.
p. 245 'cake, Canadian apples ...' *Daily Mail*, 25 May 1942, ZSL Newscuttings.
p. 245 'fifty-one tins of pink salmon ...' *Gloucestershire Echo*, 29 October 1942.
p. 245 'Her servant was twice ...' *Bristol Evening Post*, 20 January 1943.
p. 245 'using flour in the form of sausage ...' (TNA CRIM 1/1450).
p. 246 'of a suitable type ...' (TNA HO 186/1224 5647).
p. 247 'every month ...' Our Dumb Friends' League Report for 1942.
p. 248 'for so readily ...' ibid.
p. 248 'cat in pie or stew can taste ...' ibid.
p. 248 'did its best ...' ibid.
p. 248 'swill bins ...' ibid.
p. 249 'Tiger, on arrival ...' ibid.
p. 249 'He has bought ...' Goebbels, Tagebücher Band. 4 (p. 1807).
p. 250 'I feel very bitter ...' 18–19 May 1942, Klemperer, *Diaries of Viktor Klemperer 1933–1945* (pp. 496–99).

Chapter 23: The Secret War on Dogs

p. 252 'He has shown himself ...' *The Times*, 13 January 1943 (p. 5).
p. 252 'Cats queuing ...' *Muswell Hill Record* and *Friern Barnet Journal*, 15 January 1943 (p. 5).

p. 253 'Annexe ...' Soames, Mary, *Speaking for Themselves: The Personal Letters of Winston and Clementine Churchill*, Doubleday, London, 1998 (p. 471).
p. 253 fn 'Smokey' (TNA T199/274).
p. 253 'a huge dog entered ...' Sauerbruch, *Das War Mein Leben* (pp. 545–46).
p. 254 'Do certain people ...' (TNA HO 186/1023).
p. 254 'vigorous campaign ...' (TNA HO 186/2075).
p. 255 'Blood', 'Toil', 'Tunis' and 'Bizerta' ...' *Evening News*, 3 May 1943 in ZSL Newscuttings.
p. 255 'might be diverted ...' (TNA MAF 84/61).
p. 255 'With certain limited ...' ibid.
p. 255 Large amounts of ...' ibid.
p. 256 'for free painless ...' Lord President's Committee, 'The Control of Dogs', 19 February 1943 para. 11(a) (TNA MAF 84/61).
p. 256 'under review ...' (TNA MAF 128/38).
p. 256 'ten saucers of milk ...' *Daily Mirror*, 25 July 1944 (p. 26).
p. 257 'A woman at Reading ...' transcript of report in *The Star*, 22 March 1943 (TNA MAF 84/57).
p. 257 'The ruling means what ...' (TNA MAF 84/57).
p. 257 'to starve out ...' (TNA MAF 84/61).
p. 257 'partly because owners ...' Graves, *Great Days* (p. 52).
p. 257 'feeding their dogs on figs ...' ibid.
p. 258 'eight civilian trainers ...' *Our Dogs*, 17 December 1943 (p. 1223).
p. 258 'after four weeks ...' ibid.
p. 259 'Glyndr ...' (TNA WO 32/14999).
p. 259 'too difficult ...' (http://www.animalaid.org.uk/images/pdf/michael.pdf).

PART FOUR: PETS TRIUMPHANT

Chapter 24: Camp-followers

p. 265 'Dogs are found in large numbers' *The Dogs Bulletin*, spring 1943 (p. 2).
p. 266 'Dogs and cats in army camps' (TNA MAF 35/495).
p. 266 'Stray and ownerless cats ...' Army Council Instructions, 26 June 1943 (TNA MAF 35/495).

p. 266 '10,000 killer dogs ...' *Daily Sketch*, 24 April 1943 (cutting in TNA MAF 35/495).
p. 267 'fulfilled their role ...' T. J. Edwards, *Mascots and Pets of the Services* (2nd edn.) (pp. 156–7).
p. 267 'not having ...' Edwards op. cit. (p. xvi).
p. 267 'Soldiers and sailors ...' Montague, op. cit. (p. 95).
p. 268 'thousands of animals ...' Bourne, op. cit. (pp. 4–5).
p. 269 'Essex hound ...' Bourne, op. cit (pp. 75–6).
p. 269 'attractive dachshund ...' Bourne, op. cit. (p. 72).
p. 270 'she was smuggled aboard ...' (http://krrcassociation.com/swiftandbold/tich_the_desert_rat.pdf).
p. 270 'smoke a cigarette ...' *Coronet Magazine*, April 1960 (p. 111, www.oldmagazines.com).
p. 270 'Bob', 'Gyp', 'Lady' ...' 12 January 1944 (TNA WO 204/7732 7380).
p. 270 'War Diary of the Corps of Military Police ...' (TNA WO 169/13320).
p. 271 'Lost, stolen or eaten ...' Clabby, *The History of the Royal Army Veterinary Corps, 1916–1961* (p. 68).
p. 271 'a tiny terrier ...' *Tail-Wagger Magazine*, April 1944 (p. 73).
p. 272 'One soldier who was going home ...' Montague, op. cit. (p. 81).
p. 272 'asked us to come ...' ibid.

Chapter 25: Flying Pets

p. 273 'all these stories of flying kittens ...' *The Cat*, February 1943.
p. 273 'Aerodrome cats ...' ibid.
p. 273 'the alleged practice ...' *The Cat*, July 1943.
p. 274 'The rules say ...' *Tail-Wagger Magazine*, December 1942 (p. 245).
p. 274 'plug your dog's ears ...' ibid.
p. 275 'a cat called 'Windy ...' Desmond Morris, *Cat World*, Ebury Press, London, 1996 (p. 476).
p. 275 'Straddle ...' (http://www.vintagewings.ca/Vintage News/Stories).
p. 275 'Peter ...' *Tail-Wagger Magazine*, August 1944 (p. 138).
p. 275 'Antis took part ...' (http://wolfarmy.net/tag/vac lav-bozd).

p. 275 'Anthony (no. 46) …' IWM Records of the PDSA Allied Forces Mascot Club (vol. 1).
p. 275 'Flying Officer Pim (no. 670) …' ibid. (vol. 2).
p. 276 'which attach themselves …' (TNA ADM 116/5505).
p. 276 'No mascot is so popular …' *Tail-Wagger Magazine*, July 1945 (p. 129).
p. 276 'by 1943 the Charlton …' Our Dumb Friends' League Report for 1943.
p. 277 'a white Pomeranian …' (TNA MAF 35/918).
p. 277 'to an unknown USAAF …' (TNA MAF 35/918).
p. 278 'after a few hours …' (TNA MAF 35/918).
p. 278 'The two dogs were …' (TNA MAF 35/918).
p. 279 'a monkey and a honey bear …' (TNA MAF 35/918).
p. 280 'Sinbad …' (http://www.britishpathe.com/video/american-airfield-1).
p. 280 'cases so dreadful …' Our Dumb Friends' League Report for 1943.
p. 280 'discover that the tenant …' ibid.
p. 280 'that the abominable …' ibid.
p. 281 'while accompanying …' ibid.
p. 281 'He would let no one …' Smith, op. cit. (p. 54).
p. 281 'a Land Girl called Doris Adams …' ibid. (p. 64).
p. 282 'Jenny and Toni …' Heck, op. cit. (p. 98).
p. 282 'Treasure of the Gardens …' Heck, op. cit. (p. 103).
p. 283 'a wolf taking tea …' ibid. (p. 108).
p. 283 'Fantastic rumours …' Studnitz, *While Berlin Burns* (p. 140).
p. 283 'Very good were the crocodiles …' Heck, op. cit. (p. 107).

Chapter 26: The D-Day Dogs

p. 284 'bit one lady …' (TNA FO 372/4047).
p. 285 'hidey hole …' ZSL Newscuttings.
p. 285 'special duty in detecting …' (TNA MAF 35/918).
p. 286 'a dog belonging to a Flying …' Hansard, 24 February 1944.
p. 286 'Everything we see …' ibid.
p. 286 'might be executed …' (http://news.google.com/newspapers), *Milwaukee Journal*, 24 January 1944.
p. 286 'Recon went through …' (www.303rdbg.com/photo-animals.html).

p. 287 'taken on a bombing ...' (TNA MAF 35/918).
p. 288 'I carried with me ...' (http://www.pegasusarchive.org/varsity/repLuardsOwn.htm).
p. 288 'Each dog made four descents ...' ibid.
p. 288 'He subsequently endured ...' (http://www.dnw.co.uk/medals/auctionarchive).
p. 289 'I got over there a few ...' N. S. Hidgets, *Tail-Wagger Magazine*, August 1945 (p. 85).
p. 290 'black and tan Alsatian ...' (TNA MAF 35/918).
p. 291 'Retention is becoming ...' ibid.
p. 291 'The decision to tell ...' ibid.
p. 292 'Destroyed by shooting ...' ibid.
p. 292 'Somehow or other ...' Bourne, op. cit. (p. 77).
p. 292 'Might have been ...' IWM Records of the PDSA Allied Forces Mascot Club.
p. 292 'Sailor ...' *Tail-Wagger Magazine*, August 1944 (p. 146).
p. 293 'Herr General [der Hund gehtes nicht!] ...' (http://475thmpeg.memorieshop.com/CHAPTERS/FOUR/Chapter4.html).
p. 293 'you are in England now ...' ibid.
p. 293 'the dog has been flown to the UK ...' *PDSA News*, June 1945 (p. 15).

Chapter 27: Doodlebug Summer

p. 294 'a long, long line of white-faced ...' Longmate, *The Doodlebugs* (p. 149).
p. 294 'returned home the afternoon ...' ibid.
p. 295 'some time past relations ...' (TNA HO 186/2075).
p. 295 'None of our ...' 'Our Shelters Carry On', *The Dogs Bulletin*, autumn 1944 (p. 3).
p. 296 'Fortunately the animals ...' Our Dumb Friends' League Report for 1944.
p. 296 'Pussy Wake ...' Smith, op. cit. (p. 59).
p. 296 'when he heard ...' ibid.
p. 296 'When a flying bomb ...' ibid.
p. 296 'saved Corporal Witcomb ...' ibid.
p. 297 'to make an improvised ...' *The Animal World*, October 1944.
p. 297 'I have four such dog ...' ibid. (p. 74).
p. 298 'two dead cats ...' *The Animal World*, November 1944 (p. 82).

p. 298 'Over the next few ...' pps. of H. W. Atterbury (IWM Documents 13352).
p. 298 'who for five weeks has given ...' *The Cat*, October 1944 (p. 2).
p. 299 'Our dog soon realized ...' Longmate, op. cit. (p. 146).
p. 299 'Kim used to get most ...' ibid.
p. 299 'Mickie ...' ibid. (p. 147).
p. 299 'always first into the ...' ibid.
p. 299 'prick up her ears ...' ibid.
p. 299 'Benjamin ...' ibid.
p. 300 'Achtung Chimps' *Daily Mail*, 28 July 1944 in ZSL Newscuttings.
p. 300 'Binkie' Longmate, op. cit. (p. 148).
p. 300 'refused to come out ...' ibid. (pp. 148–9).
p. 300 'Ticky' *The Cat*, October 1945 (pp. 12–13).
p. 300 'one street in north London ...' *The Animal World*, November 1944 (p. 82).
p. 300 'more cattle were killed ...' (TNA MAF 52/119).
p. 301 'Highly satisfactory salvage ...' (TNA MAF 52/119).
p. 301 'large and white ...' Longmate, *Hitler's Rockets* (p. 168).
p. 302 'gave a wail of terror ...' ibid. (p. 232).
p. 302 'At the request of the Palace ...' Our Dumb Friends' League Report for 1944 (p. 27).
p. 303 'Puzzle of the Palace Kittens ...' *Daily Mirror*, 17 November 1944 (p. 6).
p. 303 'a friendly creature ...' *Evening Standard*, December 1944 in ZSL Newscuttings, 7923.
p. 304 'She is too fond of romping ...' *Evening Standard*, 18 May 1945, 7926.

Chapter 28: Finest Hour

p. 305 'perfectly comfortable ...' 'Gas Masks for Dogs' (TNA WO 188/2130).
p. 305 'Dogs with black eyes ...' 'Mine Detection Dogs', 11 August 1944 (TNA AVIA 22/871).
p. 306 'merely pretended to work ...' (TNA WO 225/1173).
p. 306 'bitches in season ...' (TNA WO 225/1173).
p. 306 'Handlers are far too kind ...' (TNA WO 225/1173).
p. 307 'some meat, oatmeal ...' *Our Dogs*, 22 December 1944 (p. 1377).

p. 308 'Animals over here ...' *PDSA News*, May 1945 (p. 3).
p. 309 'when Jet was satisfied ...' (http://jetofiada.tripod.com/Story.htm).
p. 309 'He sniffed, stepped over ...' (TNA HO 186/2572).
p. 310 'very poor co-operation ...' (TNA HO 186/2572).
p. 310 'blood marks ...' incident at Epping, 12 November 1944 (TNA 186/2572).
p. 311 'A call for silence ...' incident at Bowes Park, 14 December 1944 (TNA 186/2572).
p. 311 'jumbled cries of rescuers ...' *Daily Mirror*, 15 December 1944 (p. 8).
p. 312 'a nice red setter ...' Bourne, op. cit. (p. 165).
p. 312 'I feared his heart ...' ibid.
p. 312 'found an old lady ...' ibid.
p. 312 'after about 20 minutes work ...' (http://www.bbc.co.uk/history/ww2peopleswar/stories/60/a7011460.shtml)
p. 313 'Report of the working dogs ...' Private Papers of A. Knight (IWM Documents, 15582).
p. 314 'seemed lost without his master ...' (http://www.fdr library.marist.edu/education/resources/bio_fala.html).
p. 315 'Even Hitler set his affections ...' *Tail-Wagger Magazine*, June 1945 (p. 102).
p. 315 'Frightened apes ...' Giles MacDonogh, *Berlin*, Sinclair-Stevenson, London, 1997 (p. 142).
p. 315 'Someone came into the cellar ...' *A Woman in Berlin*, Anon, Secker & Warburg, London, 1955 (p. 185).

Epilogue: Pets Come Home

p. 321 'Today is the anniversary ...' Klemperer, op. cit. 22 May 1945 (p. 915).
p. 321 'They were led by ...' *The Times*, 11 June 1945 (p. 2).
p. 322 'impossible to understand ...' *Essex Newsman*, 26 March 1946 (p. 5).
p. 322 'Never has there ...' *The Times*, 27 February 1945 (p. 3).
p. 323 'Special Services Scheme' (TNA ADM 1/20854).
p. 323 'various canine oddments' *Dog's Bulletin*, Spring 1946 (p. 10).
p. 324 'I consider we should ...' (TNA ADM 1/20854).

p. 324 'For Valour' (TNA WO 32/14999).
p. 324 'blown up …' (TNA WO 32/14999).
p. 325 'taking random bites …' (TNA WO 32/14999).
p. 325 'native thieves …' (TNA WO 32/14999).
p. 325 'Now that hostilities …' *RAVC Journal*, November 1946 (pp. 20–21).
p. 325 'failed to recognise him' ibid. (p. 11).
p. 326 'Michael, on hearing …' (http://www.animalaid.org.uk/images/pdf/michael.pdf).
p. 328 'Khan was the one …' Rosamond Young, 'Friends in Arms' in *Chicken Soup for the Dog and Cat Lover's Soul* (ed. John T. Canfield), Health Communications Inc., FL 1999 (p. 18).
p. 328 'While the Archbishop held …' (http://www.purr-n-fur.org.uk/famous/faith.html).
p. 329 'reason was largely political …' (TNA MAF 79/10).
p. 329 'fine, fat fox …' *Horse & Hound*, 15 December 1944 (p. 7).
p. 329 'The Earl of Harrington …' *Baily's Hunting Directory 1939–1949* (p. 36).
p. 330 'most pro-Nazi …' Clabby, op. cit (p. 157). *See also* TNA FO 937/103.
p. 331 'the animals, few and far …' *The Times*, 8 January 1946 (p. 2).
p. 331 'acorns collected …' (TNA FO 943/856).
p. 333 'little is known …' 'Bravest Cat Dead', *The Times*, 1 October 1948 (p. 2).
p. 333 'Hitler', 'Timoshenko' …' *The Times*, 9 May 1955 (p. 12).
p. 333 'chose not to emphasise …' Hilda Kean (eds. Emilie Dardenne and Sophie Mesplede), *The People's War on the British Home Front: The Challenge of the Human-Animal Relationship in a Nation of Animal Lovers*, Manchester University Press, 2013.